NAMING CANADA
Stories about Canadian Place Names

Revised and Expanded Edition

Alan Rayburn

This wonderful collection of seventy-six essays explores the fascinating origin and meaning of the names of some of the towns, villages, cities, islands, mountains, and rivers that make up one of the world's largest countries. This new edition includes fifteen more essays, and updates the previous essays to include changes, corrections, and new names to the year 2000.

Discover how some of Canada's most unusual place names came to be; unearth the Aboriginal roots of names such as Miramichi, Klondike, Iqaluit, Toronto, and Ottawa; learn the origin of such playful and mellifluous names as Medicine Hat, Twillingate, Flin Flon, Cupids, or Saint-Louis-du-Ha! Ha! From Bonavista and Port au Choix in the east to Malaspina Strait and Port Alberni in the west, this book also reveals the rich Portuguese, Spanish, and Basque contributions to Canada's toponymic heritage. *Naming Canada* tells us about place names that became undesirable and had to be changed for reasons of perceived political impropriety. The former Stalin Township, for example, was renamed after Rick Hansen, the renowned Man in Motion who promoted research in spinal cord injuries. The book also discusses Canadian names that have been exported abroad, such as Quebec in England and Toronto in Australia. One new essay explores the nicknames used for Canadian places, and focuses on Hogtown as an alternative for Toronto.

This collection is the best single source, in an engaging essay format, on the origin and meaning of hundreds of Canadian place names. Alan Rayburn has had over thirty-five years of experience in researching Canada's toponymic roots and in writing about the authentic backgrounds behind thousands of names, from Toronto in the south to Tuktoyaktuk in the north, and from Labrador in the east to Juan de Fuca Strait in the west.

ALAN RAYBURN was the executive secretary of the Canadian Permanent Committee on Geographical Names form 1973 to 1987. He has written some 150 books, papers, and articles in the toponymic field, including *Naming Canada* (1994), *Place Names of Ontario* (1997), and the *Dictionary of Canadian Place Names* (1997, 1999).

Naming Canada

Stories about Canadian Place Names

Revised and Expanded Edition

ALAN RAYBURN

UNIVERSITY OF TORONTO PRESS
Toronto Buffalo London

Reprinted 1997

Revised and Expanded Edition 2001

Printed in Canada

ISBN 0-8020-4725-4 (cloth)
ISBN 0-8020-8293-9 (paper)

Printed on acid-free paper

Canadian Cataloguing in Publication Data

Rayburn, Alan
Naming Canada : stories about Canadian place names

Rev. & expanded ed.
Includes index.
ISBN 0-8020-4725-4 (bound) ISBN 0-8020-8293-9 (pbk.)

1. Names, Geographical – Canada – Miscellanea. I. Title.

FC36.R38 2001 917.1'001'4 C00-932248-5
F1004.R389 2001

University of Toronto Press acknowledges the financial assistance to its publishing program of the Canada Council for the Arts and the Ontario Arts Council.

University of Toronto Press acknowledges the financial support for its publishing activities of the Government of Canada through the Book Publishing Industry Development Program (BPIDP).

Contents

Preface

A passion for maps and places and names has its roots in my preschool days in the late 1930s outside the town of Orangeville, Ontario. During my years of attendance in the schools of Orangeville from 1939 to 1952, my love for geography and history was nurtured by numerous helpful and dedicated teachers.

The joy of studying Canadian and world history continued during several years at university, first at Waterloo College (in the mid-1950s, part of the University of Western Ontario, but now Wilfrid Laurier University), and later at the University of British Columbia, the University of Kentucky, and Université Laval. It was at the University of Kentucky that my interest in pursuing a career in place-name study received a dramatic boost. My thesis adviser, Thomas Field, was usually a mild-mannered gentleman, but he revealed considerable excitement, and sometimes animosity, when discussing the background and merits of certain place names in the state. Through his inspiration, I resolved to take up the study of toponymy, if not as a profession, at least as an avocation.

In 1957, I joined the staff of the Geographical Branch of the federal Department of Mines and Technical Surveys.

Until 1964, I was involved in land-use studies in the Maritime provinces, Québec, and Sri Lanka. In 1961, the staff serving the Canadian Permanent Committee on Geographical Names was transferred from the topographical survey division of the Surveys and Mapping Branch to the Geographical Branch. Norman Nicholson, who was both the director of the branch and the chairman of the committee, announced that a new toponymy research division would be set up in the branch, with Keith Fraser as its chief and as executive secretary of the committee.

Soon after joining the toponymy division in 1964, I undertook a place-names study of Renfrew County, a huge area of 5,000 square kilometres in the Upper Ottawa Valley. The main aims of the study were to assess the accuracy of names on topographical maps and to gather previously unrecorded names. The study revealed an error factor of 15 per cent among the official names; and for every name already in official use, at least one other new local name for another geographical feature was found to be in use.

The approach and the results of the Renfrew study appealed to the designated-names authorities in New Brunswick and Prince Edward Island, and they invited me in 1967 to undertake similar comprehensive field investigations in those provinces. The stock of approved names in New Brunswick in 1967 amounted to approximately 7,500. During an exhaustive two-year study, some 1,500 of those names were found to be incorrect as to local usage, spelling, or placement on maps; and 8,000 additional names in local usage were picked up. During the Prince Edward Island study, the stock of names was increased from 1,200 to 2,200, and 300 corrections were made. A similar study of Nova Scotia's names was undertaken by Michael Munro of the toponymy division's staff in 1973, with even more dramatic results.

In 1967, the Geographical Branch published *Geographical Names of Renfrew County* (74 pages) as part of its Geographical Paper series. Revisions of this book were privately published in 1989 and 1990. The Surveys and Mapping Branch (when the Geographical Branch had been disbanded by the Department of Energy, Mines and Resources in 1967, the toponymy division had been transferred to the Surveys and Mapping Branch) published *Geographical Names of Prince Edward Island* (135 pages) in 1973, and *Geographical Names of New Brunswick* (304 pages) in 1975.

In 1973, I was appointed executive secretary of the Canadian Permanent Committee on Geographical Names, a position I held until 1987. In that position, I worked with the committee's secretariat, and with the personnel of the federal, provincial, and territorial members of the committee in the development of national toponymic principles and procedures.

In the spring of 1983, the chairman of the committee, Jean-Paul Drolet, who was also a director of the Royal Canadian Geographical Society, proposed to the publisher and general manager of *Canadian Geographic that* a place-names column be considered for the magazine. Fortuitously, the publisher was then Keith Fraser, and he greeted the suggestion with enthusiasm.

The magazine's editor, Ross Smith, invited me to submit a few sample columns. Subsequently, the first column, entitled 'Acronyms ... Such as Kenora, Koocanusa, and Snafu Creek,' appeared in the December 1983/January 1984 issue of the magazine. By November/December 1993, sixty-one columns had been published on a variety of toponymic topics, from names pronunciation and spelling to names for royalty and governors general, and to prominent names in the news, like Meech Lake.

Fourteen more articles were published in *Canadian*

Geographic after the first sixty-one columns were selected for publication in *Naming Canada* (1994). The final fourteen columns included ones on South African names transferred to Canada, names of Basque origin in Eastern Canada, the origin and meaning of Toronto, names with a Shakespearean connection, and, for the last column, in March/April 1996, the origin and meaning of Ottawa. An additional article has been written on the nicknames of places, with special attention on the origin of Hogtown as a moniker for Toronto.

Acknowledgments

To my knowledge, the regular 'Place Names' column in *Canadian Geographic* from 1983 to 1996 was unique in that there was no similar column anywhere else in the world during the last half of the twentieth century. The original suggestion for the column was made by Jean-Paul Drolet, chairman of the Canadian Permanent Committee on Geographical Names – now the Geographic Names Board of Canada – from 1964 to 1988. As a member of the board of directors of the Royal Canadian Geographical Society, he took the idea to Keith Fraser, then executive director of the society and publisher of *Canadian Geographic*. Keith, who had been the executive secretary of the names committee in the 1960s, welcomed the idea of a names column enthusiastically. He turned the idea over to the recently appointed editor, Ross Smith. Until 1989, Ross nurtured the column by suggesting suitable ideas to write about, and by deftly trimming awkward sentences and excessive wording. His successor, Ian Darragh, and the managing editor, Eric Harris, continued to encourage the writing of articles devoted to Canada's place names. Rick Boychuk succeeded Ian as the editor in 1995. We mutually agreed to discontinue the column the following spring, when the seventy-fifth column was written on the origin of the name Ottawa. Finally, I want to acknowledge the superb advice provided by Helen Kerfoot, my successor as executive secretary of the committee from 1987 to 1998.

NAMING CANADA
Stories about Canadian Place Names

Revised and Expanded Edition

Introduction

The 1994 edition of *Naming Canada* comprised the sixty-one 'Place Names' columns first published in *Canadian Geographic* from December 1983/January 1984 to November/December 1993. This 2001 edition contains those sixty-one columns, revised and updated. In addition, the fourteen columns published in the magazine between January/February 1994 and May/June 1996 are included in this edition, also revised and updated. A new essay on nicknames of places, focusing on Hogtown for Toronto, has been added.

Several corrections and changes have been made to reflect information provided by the magazine's readers. Some new details, dropped from the original articles because of space limitations, or having come to light since publication in the magazine, have been added to some of the essays.

The idea for bringing the columns together in a single volume was initiated in 1993 by Bill Harnum, vice-president of scholarly publishing at the University of Toronto Press. The initial arrangement of the essays into particular themes was undertaken by Laura Macleod. Subsequently, Gerry Hallowell, Rob Ferguson, and Beverley Beetham Endersby completed the editing of the texts for publication in 1994. This new edition has been arranged and edited by Jill McConkey, Frances Mundy, and Beverley Beetham Endersby.

The essays are grouped into eight broad themes.

1 *Looking at Canada's Places from the Atlantic to the Pacific, and to the Arctic.* The article on the origin of the name Canada, and the amusing alternatives proposed for the name of the new nation in 1867, was the thirty-seventh article published in the magazine, but it is the obvious lead-off essay in this collection. The article about the suppression of the title 'Dominion' was a natural second essay. The remaining six articles of the group deal with the highest points in each of the provinces and territories, Canada's national parks, my favourite 24 names from coast to coast, Canada's federal electoral ridings, nicknames of Canadian places, and the farthest points on the compass.

The declaration in the farthest-points article that there was an unnamed point farther north than Cape Columbia persuaded the Department of National Defence's Mapping and Charting Establishment to undertake precise measurements. The point was found to be 242 metres farther north than the designated most northerly site of Cape Columbia. The Canadian Permanent Committee on Geographical Names decided in 1992 to shift Cape Columbia to the point. The article on the federal electoral districts encouraged the staff of Elections Canada to set up an advisory committee for the selection of names of the new 301 ridings based on the 1991 census, and I was invited to serve on the committee. The story about nicknames examines the labels and monikers given to Canadian places, with special attention devoted to Hogtown, the widely known reference to Toronto. The article on the highest points in each jurisdiction guided sexagenarian climbers Clifford and Ruth Holtz of Oshawa, Ont., in their ascents in 1990 of each of the eleven highest points in the country – Québec and Newfoundland share the same highest elevation, but with different names. And one of the points of the Y-shaped summit may be nearly a metre higher in Newfoundland.

2 *Scrutinizing Political Issues and Language Problems.* Eleven of the essays concern a number of language issues and a variety of political problems involving names. In French, words like 'Québécois' and 'Acadien' are called *gentilés*, but no suitable English word for them has evolved to describe them. *Gentilés* are widely used all across Canada, from Haligonian to Winnipegger and on to Vancouverite. Suggestions on what I believed were the accepted pronunciations of several names aroused a couple of readers, who forcefully pointed out my errors. Although Canada does not have too many confusing sound-alikes and look-alikes, there are bothersome ones like St. John's/Saint John, Kenora/Canora, two major Churchill Rivers, and too many places called Windsor. The misspelling of place names is quite common for places like St. Catharines and Edmundston, and it doesn't help when there is more than one official spelling, such as Timiskaming/Témiscaming/Témiscamingue, and Okanagan north of the 49th parallel, Okanogan south of it. In another essay, actual mistakes, such as Cedoux evolving from Ledoux, and Barry's Bay from Byers Bay, are examined. Several names are acronyms, that is, created from parts of two or more words. Among them are Arvida, Estevan, and Tarfu Creek. The handling of the possessive apostrophe in names has had a mixed history over the past hundred years, with places like St. John's preferring it, and Smiths Falls ignoring it. The essay on merging municipalities reveals the problems encountered when new municipalities require their own identity, such as Thunder Bay, Clarington, Miramichi, and Cambridge.

Political correctness forced the change of Berlin to Kitchener, and the dropping of Mount Stalin, but name purifiers have been unable to divest Swastika of its 'good luck' identity. Sensitivity resulted in the banishment of Belly River and Foul Bay, but Crotch Lake and Bastard Township have survived attack. When former prime minister John Diefenbaker requested background material in 1976 relating to the public clamour to change Mount Eisenhower back to Castle Mountain, I assumed he would rise in the House of Commons to condemn the action taken in 1946 by Mackenzie King to assign a foreigner's name to one of Canada's most impressive mountains. However, he roundly condemned any attempt to sully the honour of the Supreme Commander of the Second World War forces, and Liberal minister Alistair Gillespie, who was responsible for the names committee, fully agreed with him. But three years later, during Joe Clark's brief government in 1979, we got our Castle Mountain back.

3 *Receiving Names from Abroad and Exporting Names.* Great numbers of Canada's names first saw the light of day in foreign lands. Eleven essays concern the transfer of names from abroad, and a twelfth in this group of articles is about Canada's export of names to other countries. Athens, London, and Paris recall great world capitals. Canaan, Goshen, and Damascus owe their existence to biblical references. Zealandia and Australia Gulch are among many names drawn from Canada's Commonwealth partners down under. Portugal and Spain were leaders among early explorers, so it is not surprising to find names like Bonavista and Juan de Fuca Strait on our east and west coasts. Basque whalers and fishers left a legacy of more than 300 names on Canada's east coast, primarily around the shores of the Gulf of St. Lawrence and along the estuary of the St. Lawrence River. Hispanic names from the United States and Latin

America include Almonte, Spanish River, Toledo, Valparaiso, Eldorado, and Bonanza Creek. Hard-fought victories in the Crimean Peninsula by British and French forces introduced names like Alma, Inkerman, and Canrobert to Canada's toponymic fabric. The Anglo-Boer War at the beginning of the twentieth century provided several names for Canadian places, including Ladysmith, Mafeking, Redvers, Mount Rhodes, and Kimberley. First World War battles and victories in northern France and in Belgium provided more than 125 names for Canadian places and mountains, examples being Vimy Ridge, St-Quentin, and Messines. Names reflecting heaven, peace, and harmony are widely affixed to Canadian places, from Paradise in Newfoundland to Utopia in Ontario, and Sointula in British Columbia. Canadian names like Toronto and Ontario have been transferred to other countries, especially the United States and Australia, and even names like Québec have been given to places in Britain and France.

4 *Revealing Special Characteristics of Place Names and Generic Terms.* Ten essays are devoted to special characteristics of Canadian place names and to generic terminology. The lead-off article reveals that Toronto has the largest population of any incorporated 'city' in Canada, and Greenwood, B.C., has the least population. In terms of area, Timmins is Canada's largest, with Outremont, Qué., the smallest. An essay is devoted to the cities, towns, villages, and townships of Ontario, where the 815 municipalities that existed in 1997 have been reduced to fewer than 450 four years later. The wide use of 'butte' for a flat-topped hill and the frequent occurrence of 'coulee' for a steep-sided valley are legacies of the French voyageurs and explorers who first investigated the resources of Western Canada. In Eastern Canada, there is a fascinating variety of terminology in names like Pinch-

gut Tickle, Pocomoonshine Deadwater, MacLean Barachois, Mersereau Bogan, Pickards Padou, Framboise Intervale, and Meetinghouse Rips. Names like Stratford, Avon River, Mount Romeo, Macbeth Icefield, and Hamlet Township are among many names derived from the rich depository of English literature penned by William Shakespeare. Christmas is a joyous time, so it is not surprising to find names like St. Nicholas Peak, Tiny Tim Lake, and Yule Rock celebrating that season. Names like Lovers Cove, Heart's Delight, and Bow and Arrow Shoal inspired the writing of the essay on features with a theme relating to the playfulness and romance associated with Valentine's Day. In a contrasting theme, there are hundreds of names dedicated to Satan and his traditional homes of hell and Hades. Ominous examples include Backside of Hell Cove, Devils Elbow Rapids, and Ogre Mountain. An essay on telenaming the landscape draws attention to names like Signal Hill, Marconi Towers, Telegraph Creek, Satellite Hill, and Landsat Island. The final essay of this series concerns the much misunderstood term 'haha' that occurs in several names in Eastern Canada, the best example being the village of Saint-Louis-du-Ha! Ha! Even though the term means 'an unexpected barrier,' the vast majority of people will probably continue to believe it was introduced in names for comic relief.

5 *Adopting Names of Native Origin and Acknowledging Names Used by Indigenous People.* Names of Native origin are distinguishing marks of a country. The first of eight essays draws attention to the strength and rustic beauty of names like Athabasca, Muskoka, and Keremeos. The second explores the trend to acknowledge the names actually used by the Native residents of places like Iqaluit, the largest settlement in Nunavut, and Kuujjuaq in Northern Québec, formerly Frobisher Bay and Fort

Chimo, respectively. The next six essays celebrate individual names with Native roots: Miramichi, Saskatoon, Medicine Hat, Mississippi River, Canso, and Nanaimo. To me, all six add an extra patina of beauty to the regions in which they are located. The last article of the series explores names like Moose Jaw, Mooseland, and Moose Bath Pond, which owe their origins to the word *moos*, from the description by Algonkian-speaking tribes of the browsing habits of the hulking animal occupying the Canadian wilderness from coast to coast.

6 *Examining the Names of Particular Places.* The eight essays in this next grouping are primarily about names of single places, with the first one about Cupids, Nfld., where the oldest British colony in present-day Canada was established in 1610. Acadia is the mystical home of a quarter-million people, who proudly proclaim their common heritage, but the name's roots are uncertain, and the area both historically and currently constituting Acadia widely differs among Acadians. Montréal is the most vibrant city in Canada, but the derivation of its name from the adjacent 223-metre Mont Royal has been long disputed. The next article traces the roots of a name on every Canadian's lips in the late 1980s – Meech Lake – and examines the dual names of the next lake to the northwest, known as Harrington Lake in English, Lac Mousseau in French. Ottawa, Canada's capital, has a name rooted in the fur trade between an Aboriginal tribe and the French in the mid-1600s. The name Toronto may be traced to the practice of Natives catching fish in weirs embedded beneath the water between Lake Simcoe and Lake Couchiching. The story of the origin of Flin Flon seems more fanciful than real, but it's an expressive name for a mining city set among Manitoba's rugged rocks. Robert Service's ballad of the cremation of Sam McGee on the shores of the Yukon's Lake Laberge, and

correspondence with the man it is named for, Michel Lebarge himself, provide an interesting tale of naming in Canada's Yukon.

7 *Observing Selected Names in Particular Regions.* Six of the essays deal with several names in particular regions. The lead-off article examines the incredible 4¾-year voyage of George Vancouver to Canada's west coast in the late 1700s, and the naming of features from the Strait of Georgia to Alaska's Cook Inlet. During the same period, Alexander Mackenzie undertook two arduous journeys, one to the mouth of the country's longest river, subsequently named for him, the second to the shores of the Pacific. Eleven geographical features honour Mackenzie's prodigious accomplishments. Calgary provides the focus for an essay about features associated with the 1988 Winter Olympics held there. The Thousand Islands comprise 1,149 identifiable pieces of rock and land surrounded by the waters of the upper St. Lawrence River, but only 367 in both Ontario and New York have official names. Downriver at Cornwall, the construction of huge dams in the 1950s resulted in the inundation of 8,000 hectares of land; the relocation of eight villages; and the creation of two new towns, Long Sault and Ingleside. The naming of the historic passes on the Great Divide of the Rocky Mountains is the subject of the last article of this series of essays.

8 *Commemorating Prominent Individuals and Honouring Certain Family Names.* The final twelve essays concern features named for noted individuals, or for particular surnames. The revered Guy Carleton, Baron Dorchester, has been honoured in several names in Eastern Canada. There are 218 features with the name Fraser in Canada, the most famous being British Columbia's Fraser River. George Dawson, a brilliant scientist and scholar who overcame physical adversity, named many features him-

self, and in his honour at least 25 features were named for him. No single individual has had more features – more than 300 – named for her in Canada than Queen Victoria, beginning with the capital cities of both British Columbia and Saskatchewan. Four Queen Elizabeths have been honoured in the names of islands, mountains, roads, and public buildings. The Queen's eldest son, the Prince of Wales, had several features named for him during his visit to Canada in 1860. When his eldest son, also the Prince of Wales, visited Canada in 1919 a number of features were named in his honour. In 1979, the museum of Native artifacts in Yellowknife was named for the present Prince of Wales. The nineteenth governor general, Georges Vanier, was a respected statesman for whom several features, including cities in Ontario and Quebec, were named. In 1927 a group of peaks in British Columbia's Cariboo Mountains was chosen to be named after prime ministers. Four were named that year after for Laurier, Abbott, Thompson, and Bowell. During the next forty-five years peaks were selected to honour King, Meighen, Bennett, St-Laurent, and Pearson. The naming of a mountain for Roland Michener and a lake for Jules Léger, Canada's twentieth and twenty-first governors general, reflects the great respect these public servants had among their countrymen. The most honoured Canadian in the naming of places is Sir Wilfrid Laurier, with at least 37 geographical features as well as a number of public institutions having been named for him. President John F. Kennedy had a mountain named for him in 1964, and the following year it was climbed by his brother Robert. The final essay is about features across the country named Smith, the commonest surname in several countries in the world. The inspiration to write this article came to me during the summer of 1984, when, while visiting the editor of *Canadian Geographic*, Ross Smith, I thoughtlessly observed that consideration had once been given to changing the magical name Medicine Hat to the dull and uninspiring Smithville.

Here's to all those Smiths, Frasers, Poiriers, Macdonalds, Cartiers, Mercredis, Englebrechts, and millions of other Canadians of many races and cultures who have made this great land such a memorable experience, toponymically speaking. Thanks for *Naming Canada.*

Looking at Canada's Places from the Atlantic to the Pacific, and to the Arctic

Canada: A Native Name from the Land

'That's the way to Canada,' Jacques Cartier's young Aboriginal companions, Taignoagny and Domagaya, shouted as the explorer neared Î'le d'Anticosti on the St. Lawrence River on 17 August 1535. Cartier turned his ship up the 'grande rivière de Hochelaga,' also called 'rivière de Canada,' and by 7 September was within 100 kilometres of Stadacona (site of present-day Québec City) in Chief Donnacona's territory, called Canada.

Donnacona, of a St. Lawrence Iroquoian-speaking tribe, was the father of Cartier's two Native friends, having allowed his sons to accompany the explorer to France the previous year. He and Cartier had met near what is now the town of Gaspé. Cartier later learned that Canada extended another 50 kilometres beyond Stadacona (the limits today would be Baie-Saint-Paul on the east, Portneuf on the west, and the south shore of the St. Lawrence River).

In his report of the 1535–6 voyage, Cartier appended a list of the local St. Lawrence Iroquoian-language words. He noted that *kanata* meant 'town,' interpreted subsequently as meaning a cluster of dwellings, with the lands ruled by Donnacona being a series of towns. André Thevet, a contemporary French explorer, wrote that the name meant 'land.'

The words for town or village in various Iroquoian languages are similar. The Mohawk use *nekantaa,* the Onondaga *ganataje,* and the Seneca *iennekanandaa.*

So the origin and meaning of 'Canada' would appear to be beyond dispute. However, that has not prevented a proliferation of other stories, some of tenuous credibility.

It may have been Father Louis Hennepin who first observed the similarity of the Spanish *aca nada* ('here nothing') to Canada. He reported in 1698 that early Spanish explorers were disappointed in not finding gold and other riches in Canada, and frequently made that derisive declaration. It is said that Native people picked up the phrase and passed it to Cartier as the name of their country. The historian Charlevoix mentioned it as an old tradition in his history of New France, published in 1744.

Some writers give the credit to the Portuguese, who also are said to have exclaimed in disgust *cà nada* ('here nothing').

In the 1880s, Marshall Elliott, an American writer, was convinced that a Spanish or Portuguese origin was correct and did much to cast doubt on the name's Native provenance.

In 1760, Thomas Jefferys, a British map maker, published a geography in which he suggested the name was derived from a Native language and that it meant 'mouth of the country,' an allusion to the mouth of the St. Lawrence River.

A writer in an issue of *Nova Scotia Magazine* of 1779 offered that a Frenchman called Cane had attempted to found a colony but failed, leaving only his name for the territory.

The most outrageous suggestion came from a writer in

the *Kingston Gazette* in 1811. He said the origin could be traced to the first French settlers demanding a 'can a day' of spruce beer from the intendant, the colonial administrator.

In 1861, Rev. B. Davies, an English philologist, put forward an Oriental origin, noting the similarity of the name Canada with Canara or Carnata in the south of India. Such speculation was not based on a thread of historical or linguistic evidence.

Other suggestions have included the Latin word *candida*, the Sanskrit *kanada*, and the Spanish *cañada*, all with no more evidence than the mere coincidence of similar syllables.

The Dictionary of Canadianisms (1967) states that the etymology of Canada is by no means clearly established, which imparts a measure of respect to the many fanciful suggestions that have been proposed to discredit the information from Jacques Cartier in the sixteenth century.

Writers of the twentieth century, such as Francis Parkman (*Pioneers of New France*, 1905), Mark Orkin (*Speaking Canadian English*, 1970), and William B. Hamilton (*The Macmillan Book of Canadian Place Names*, 1978), have concluded that the weight of opinion favours the Native source as reported by Cartier.

Maps made in France soon after Cartier's return in 1536 assigned the name Canada to a vast territory north of Île d'Anticosti. An exception was a map made for Nicolas Vallard in 1547, where the name is inscribed in the area of the present city of Québec.

Pierre Desceliers, an eminent cartographer near Dieppe, produced several maps of eastern North America in the mid-1500s. His 1550 map shows the name Canada four times. Near the head of the Saguenay it is written in large letters once, and in smaller letters twice. It is shown again on the south shore of the St. Lawrence

River opposite the mouth of the Saguenay. Why the repetition? Perhaps he was told that the small lettering did not give sufficient prominence to Donnacona's territory, and was persuaded to inscribe the name in larger letters.

For more than two centuries, from the 1550s to the late 1700s, Canada was used as a popular alternative to New France, at least to that part bordering on the shores of the 'grande rivière de Canada,' the St. Lawrence. It was not until 1791 that it was given official status when the Constitutional Act created the two jurisdictions of Upper Canada and Lower Canada. The proclamation of the Act of Union in 1841 amalgamated the two regions into a single province of Canada, divided into Canada East and Canada West. But Upper Canada and Lower Canada remained in common use, even in debates in the Legislative Assembly of Canada.

In the 1860s, when union with the Atlantic provinces was in the offing, proposals for a suitable name for the confederation included not only Canada, but more than thirty others. Among those most frequently repeated were Tuponia or Tupona (an acronym for The United Provinces of North America) and Efisga (England, France, Ireland, Scotland, Germany, Aboriginal). Other suggestions included Acadia, Albertland, Albertoria, Albionara, Albona, Alexandrina, Aquilonia, Borealia, Britannica, Cabotia, Canadensia, Colonia, Hochelaga, Laurentia, Mesopelagia, New Albion, Niagarentia, Norland, Superior, Transatlantica, Transylvania, Ursalia, Vesperia, Victorialand, and Victorialia.

When Thomas D'Arcy McGee addressed the Legislative Assembly on 9 February 1865 on the subject of the proposed union, he noted such suggestions as Tuponia and Hochelaga, and remarked whimsically: 'Now I would ask any Hon. Member of the House how he would feel if he woke up some fine morning and found himself, instead of a Canadian, a Tuponian or Hochelagander?'

When the British North America Act was declared on 1 July 1867, it announced that Canada would be the name of the new Dominion. *The Canadian Encyclopedia* notes: 'the name Canada reveals strength, generates pride and reflects much of the land's rugged character and its resourceful people.'

How Canada Lost Its 'Dominion'

When many of us were growing up, we spoke about living in the Dominion of Canada, heard the word *Dominion* almost daily in the news and in references to our nation, and proudly celebrated Dominion Day every First of July. When the prime minister and premiers met, it was called a Dominion–Provincial Conference. In school we learned about the indomitable Dominion Land Surveyors who carved up the West, and on the classroom wall there was a distorted Mercator map of the world with a huge, bright pink land mass labelled in bold letters DOMINION OF CANADA. But we rarely hear or see Dominion any more. Why?

I did not notice the erosion of the word from our political lexicon until it was officially dropped from the name of the 1 July holiday in October 1982. About that time, I was asked by the United Nations to confirm the official long and short names of our country. I assumed the long title was Dominion of Canada, and the short was simply Canada. I was wrong. External Affairs declared that Canada alone was official as both the long and the short name.

Then early in 1990, an article on Canada's future in *Maclean's* magazine said that Canada 'is not, in any official usage, a kingdom, a commonwealth, republic or federation, much less a union. Nor is it anymore, in common government parlance, a dominion.' This set me off on an odyssey to discover when we had officially ceased to be the Dominion of Canada. To my surprise, I found the title has not been officially dropped; it has only been suppressed, with *federal*, *national*, and *central* substituted as adjectives, and *Canada*, *nation*, and *country* used to replace the noun.

At one time, Canada's public servants included the Dominion Fire Commissioner, Dominion Stone Carver (Parliament Buildings sculptor), Dominion Archivist, Dominion Astronomer, Dominion Statistician, and Dominion Cerealist. None of those titles can be found in the Government of Canada telephone directory today. In 1994 three titles remained in use: Dominion Geodesist, Dominion Hydrographer, and Dominion Carillonneur. By the end of the twentieth century, Dominion Geodesist had been replaced by Director of Geodetic Surveys, and Dominion Hydrographer had all but disappeared in favour of Director General of the Canadian Hydrographic Service, leaving the single title of Dominion Carillonneur to carry on into the new century. The only title in French with *Dominion* still retained is that of *Carillonneur du Dominion*.

The word *dominion* had been used as early as 1764, when George III was encouraged by the Lord Commissioners of Trade and Plantations to obtain 'accurate surveys of all your Majesty's North American dominions.' Historians trace the origin of the title Dominion of Canada in the constitution to Sir Samuel Leonard Tilley, one of the Fathers of Confederation, from New Brunswick. While assembled at the London Conference in December 1866, the delegates from the provinces of Canada, New Brunswick, and Nova Scotia were pondering what to call the proposed federation. Kingdom of Canada had been considered but was judged pretentious and possibly offensive to Americans. Besides, Queen Victoria did not like it.

One morning, Tilley reached for his bible on awakening and read: 'He shall have dominion also from sea unto sea, and from the river unto the ends of the earth.' The appropriateness of the word struck him at once. Here was a term to fulfil the hopes of creating a great nation from the Atlantic to the Pacific, and from the mighty St. Lawrence to the vast northern reaches. The other delegates later agreed, and Lord Carnarvon, the colonial secretary and chairman of the conference, persuaded the Queen and the British prime minister, Lord Derby, to accept the title Dominion of Canada.

On 29 March 1867, the British North America Act received royal assent, and a new word appeared in the world's political vocabulary. The act stated that 'the Provinces of Canada, Nova Scotia, and New Brunswick shall form and be One Dominion under the Name of Canada; and on and after that Day those Three Provinces shall form and be One Dominion under that Name accordingly.'

When the Canadian constitution was patriated in 1982, the entire British North America Act was incorporated into it as the Constitution Act, 1867. So the word *Dominion* continues to be a part of the official title of this country (although its *legal* name is strictly Canada).

Until the Second World War, Canadian politicians, historians, and writers comfortably spoke and wrote about the Dominion. But during and immediately after the war, considerable tensions over Québec's relationship to Ottawa developed between Premier Maurice Duplessis and Prime Minister Mackenzie King. With the appointment of Louis St-Laurent as minister of external affairs in 1946, and his becoming prime minister two years later, the tensions were eased by quietly dropping references to the Dominion, viewed by Duplessis as an oppressive word implying Québec's subservience to the government in Ottawa.

By the 1960s, many titles of agencies in Ottawa were stripped of the word *Dominion.* The Dominion Bureau of Statistics became Statistics Canada; Dominion experimental farms became research stations. As early as 1948, the magazine *Saturday Night* referred to 'federal ministers' and 'national capital,' using *Dominion* only in relation to official functions. And by 1951, *Queen's Quarterly* was writing of 'federal–provincial conferences.'

From the 1940s to the 1970s, various private members' bills to change Dominion Day to Canada Day were 'talked out' in the House of Commons. Regardless, almost all references in the media and by the federal government after 1972 were to Canada Day.

Then, in May 1980, Hal Herbert, Liberal MP for Vaudreuil, introduced a private member's bill to replace Dominion Day with Canada Day. Many MPs expected nothing would come of it, but on 9 July 1982, the bill was given second reading and referred to the Committee of the Whole House. Although a quorum was not in the House, the call for third reading was made immediately. When no objections were heard, the bill was carried and the House adjourned.

The bill was then sent to the Senate, where a vigorous debate took place in October. Although a stout defence of Dominion Day was advanced by Senator Eugene Forsey and the Monarchist League of Canada, the Senate, aware of a Gallup poll indicating 70 per cent of Canadians favoured the change, approved the bill without a recorded vote.

On 1 July 1989, Michael Valpy, a *Globe and Mail* columnist, assailed Canadian parliamentarians as 'vandals' and 'milquetoast stewards,' but it seems that most Canadians are content with our national holiday being called Canada Day.

Across the land, however, remnants of Canada's title still flourish: Dominion Square in Montréal; the community of Dominion, N.S.; the village of Dominion City, Man.; New Dominion, P.E.I.; Dominionville, Ont.; and several creeks, rivers, lakes, and mountains called Dominion. The southwestern Ontario town of Ridgetown is served by the *Ridgetown Dominion* weekly newspaper, and many businesses, from Dominion Cellulose Ltd. to Toronto-Dominion Bank, still proudly celebrate the historic title of our country.

As we enter the twenty-first century, Dominion remains in the title of our country because section 3 of the British North America Act – to 'form and be one Dominion' – is still in Canada's constitution.

Canada: The North, South, East, and West of It

The most easterly land of Canada – indeed of North America – is Cape Spear, which juts into the Atlantic at a longitude of 50°37'W.

Not long after the voyage of the Portuguese explorer João Fernandes in 1498 (partly in company with John Cabot), the names *Cauo de la spera* and *Riuo de la spera* appeared on some maps published between 1505 and 1508. These Portuguese names refer to the idea of 'waiting.' The small bay on the northwest side of the cape was an important rendezvous for the Grand Banks fishery in the early 1500s.

In the 1540s, the name of the point was rendered on maps in French as *Cap d'Espoir* (meaning 'cape of hope'), and later as *Cap d'Espérance.* This was rendered into

English as Cape Spear and that became the established form on British Admiralty charts.

The second Portuguese name, *Riuo de la spera*, became Cape Bay on early English charts, but was changed to Spear Bay in the 1700s. In 1981, the name Cape Bay was restored; Spear Bay Brook has remained for a small stream flowing into the bay.

The most southerly land feature of Canada is Middle Island in Lake Erie at latitude 41°41′N. (So much for thinking of Canada as entirely north of the 49th parallel!) Middle Island is at the same latitude as the Klamath Mountains of northern California; Great Salt Lake, Utah; Des Moines, Iowa; and Chicago, Ill. The island was purchased from its American owners in 1999 and is being administered by Point Pelee National Park.

Middle Island itself lies in the middle of the Lake Erie shipping channel, almost halfway between Pelee Island (Ont.) on the north, and Kelleys Island (Ohio) on the south.

Pelee Island is our most southerly occupied land. The name of the island is derived from nearby Point Pelee, itself named by the French because its east side was found to be devoid of trees, and thus *pelée* ('bare' or 'peeled').

Father Pierre Charlevoix, who took a journey along the shore of Lake Erie in 1721, noted in his journal, published in Paris in 1744, that Point Pelee was 'well enough wooded on the west side, but on the east side it has a sandy soil with only small red cedars of inconsiderable quantity.'

On looking at some maps of Canada, some people may conclude that the country's most westerly point is Frederick Island, the farthest west of the Queen Charlotte Islands. The longitude of this island is 133°11′W, but it is nearly eight degrees east of the Yukon–Alaska boundary at 141°. The southern point of this boundary is Mount St. Elias; its northern mainland limit is Demarcation Point.

Mount St. Elias was first noted on 16 July 1741 by the Danish-born explorer Vitus Bering during a voyage of discovery on behalf of the Russian government. On St. Elias Day, four days later, he named Cape St. Elias on Kayak Island in present-day Alaska. Subsequently, map makers assigned the cape's name to the mountain sighted but left unnamed by Bering, and to the highest range of mountains in Canada.

Demarcation Point, the most northwesterly point of Canada, was mentioned by Sir John Franklin in his *Narrative* of 1828. He noted it as having been so named 'from its being situated in longitude 141°W, the boundary between the British and Russian dominions on the northern coast of Canada.'

The name of the most northerly point in Canada has been debated for a number of years.

During his expedition to northern Ellesmere Island in 1875–6, Sir George Nares applied the name Cape Columbia (from the poetic name of the United States) to the most northerly extension of land at 83°06′N. At the same time he named an eastern extension of the same body of land Cape Aldrich for his first lieutenant, Pelham Aldrich, who had established that the vicinity of the capes was the most northerly land of the Arctic Archipelago.

In 1911, the Geographic Board published several decisions on names. Its approval of Cape Columbia includes the statement: 'Name given by Nares, as it is the most northerly point of North America.'

Moira Dunbar and Keith Greenaway, in their *Arctic Canada from the Air* (1956), stated without qualification that Cape Aldrich is the northernmost point. They based this judgment on their interpretation of air photos.

In his *North of Latitude Eighty* (1974), Geoffrey Hattersley-Smith also asserted that Cape Aldrich is the most northern point by a few hundred metres, but noted that 'Aldrich himself accorded the distinction to Cape Columbia.' In the 1950s and 1960s, Dr. Hattersley-Smith raised with the names authority many questions about various names of northern Ellesmere Island, but he did not draw attention to any problems relating to the most northerly named point. In 1967, geographer Jim Lotz suggested naming the end of the whole 15-kilometre-wide body of land, including the two named capes, as 'Centennial Promontory.' No action was taken on this proposal.

The most recent representation on topographical maps of the area of Cape Columbia is the 1970 edition of the 1:500,000 scale map. It and the 1967 edition of the 1:250,000 scale map reveal a small point extending farther north than the points identified as Cape Columbia and Cape Aldrich.

When this article appeared in the Feb./Mar. 1987 issue of *Canadian Geographic,* it contained my suggestion that Cape Columbia could continue to be recognized as the most northerly land of Canada, but that the most northerly point of the cape be given a uniquely Canadian name.

In the summer of 1987, the Mapping and Charting Establishment of the Department of National Defence surveyed the area of Cape Columbia and determined that the small point was indeed the most northerly piece of land, being 242 metres north of the site identified as Cape Columbia.

In 1992, the Canadian Permanent Committee on Geographical Names decided that the small point, at the latitude of 83°06′41.35″, would henceforth be known as Cape Columbia.

High Points across the Land

The highest mountain in Canada is in the St. Elias Mountains in the Yukon. At 5,959 metres, Mount Logan is second only to Mount McKinley (6,194 metres) in Alaska as the highest elevation in North America. In 1890, I.C. Russell of the U.S. Geological Survey, while undertaking a survey in the St. Elias Mountains, named it for Sir William Edmond Logan (1798–1875). Born in Montréal and educated in Scotland, Logan founded the Geological Survey of Canada in 1842. On 4 October 2000, Prime Minister Jean Chrétien stated that the mountain was to be renamed for former prime minister Pierre Elliott Trudeau, who had died a week earlier. However, after two weeks of considerable negative reaction to the proposal in the media, Heritage Minister Sheila Copps announced that another, unnamed feature would be selected to honour Trudeau's memory.

The St. Elias Mountains were named in the late 1800s after Mount St. Elias, a name in turn derived from Cape St. Elias in Alaska, named by the Danish explorer Vitus Bering on St. Elias Day in 1741. Mount St. Elias is on the western boundary of Canada and, at 5,489 metres, is our second-highest mountain. In fact, the next sixteen highest mountains in Canada are all in the St. Elias Mountains, all but one entirely in the Yukon. Fairweather Mountain (at 4,663 metres, the eighth-highest in Canada) straddles the B.C.–Alaska boundary. It was named by Capt. James Cook during his historic voyage along the west coast in 1778.

The highest mountain entirely within British Columbia is Mount Waddington (4,016 metres) in the Coast Mountains. It was named in 1918 by the Geographic Board of Canada for Alfred Waddington, who from 1858 to 1872 promoted the dream of a transcontinental route to the Pacific via Bute Inlet, the head of which is 60 kilometres south of the mountain.

Alberta's highest mountain is Mount Columbia (3,747 metres), located on the boundary with British Columbia in the Rocky Mountains. It received its name in 1899 from the Columbia River, itself named in 1792 by Capt. Robert Gray, an American, for his vessel, the *Columbia*. But Mount Columbia is not the highest peak in the Rockies; that claim belongs to British Columbia's Mount Robson, whose elevation is 3,954 metres, but still only twentieth highest in Canada. The origin of the Robson name is uncertain. One story is that it may have been named for Colin Robertson (1783–1842), a Hudson's Bay Company trader.

The highest point in Saskatchewan is an officially unnamed elevation in the Cypress Hills, which rises to 1,392 metres near the Alberta border. (The Cypress Hills rise even higher, to 1,465 metres, farther west in Alberta.) Early French voyageurs identified the jackpine as a *cyprès*, and this was rendered as 'Cypress' on the map of the Palliser report of 1857–60.

The highest crest in Manitoba is Baldy Mountain, at 832 metres. It is located in Duck Mountain Provincial Park, 60 kilometres northwest of Dauphin.

The highest peak in the Northwest Territories is officially unnamed, although referred to as 'Mount Nirvana' in the alpine literature. Called that in 1965 by William J. Buckingham, a mountain climber, it rises to 2,773 metres. The mountain is in the Ragged Range, southwest of the South Nahanni River. The range was named in 1960 by Hugh S. Bostock for the characteristic ruggedness of its summits. At an elevation of 2,764 metres, Mount Sir James MacBrien, 30 kilometres to the north of Mount Nirvana, is usually given the honour of being the highest in the Northwest Territories, but it is short by 9 metres. A native of Port Perry, Ont., MacBrien (1878–1938) was a distinguished soldier and commissioner of the RCMP, 1931–8.

The highest point in Nunavut is Barbeau Peak, named in 1969 for the distinguished anthropologist and folklorist Marius Barbeau (1883–1969).

The highest spot in Ontario, 95 kilometres north of Sudbury, was determined only in 1972 when Ishpatina Ridge, with an elevation of 693 metres, was found to be 28 metres higher than Ogidaki Mountain. The latter, near Sault Ste. Marie, was declared in 1966 to be the highest elevation. Before that, Tip Top Mountain near Lake Superior, at 640 metres, was considered the highest. In the Ojibwa language, *Ishpatina* means 'high hill' and *Ogidaki* means 'high ground.'

The highest summit in Québec, at 1,652 metres, is Mont D'Iberville in the Torngat Mountains. This name was given in 1971 by the Commission de toponymie du Québec for Pierre Le Moyne D'Iberville (1661–1706), who led many ruthless expeditions in North America, including a destructive rampage in St. John's and the Avalon Peninsula of Newfoundland in 1696–7. In 1981, the Newfoundland Geographical Names Board ascertained that the peak so named was on the Québec–Labrador boundary, and decided that a name more suitable to that region's history should be assigned to the peak. The board gave it the name Mount Caubvick, in honour of an Inuit woman, who, with four other Inuit, accompanied George Cartwright, a trader on the Labrador coast, to England in 1772. So it is no coincidence that the heights of the highest points in Québec and Newfoundland are identical: they are the same feature, but with two quite different names.

On the island of Newfoundland, the highest elevation, at 814 metres, is an unnamed peak in the Lewis Hills in the Long Range Mountains. The summit is 40 kilometres southwest of Corner Brook, and only 6 kilometres from the Gulf of St. Lawrence.

The highest land in Nova Scotia is White Hill, a treeless peak with an elevation of 532 metres on the North Barren of the Cape Breton Highlands, 17 kilometres west of Ingonish.

Prince Edward Island's highest elevation is only 142 metres above sea level, and is located in Lot 67, Queens County. The community on the hill is rather oddly named, being composed of two words suggesting a place lower than its surroundings: Glen Valley.

In 1899, W.F. Ganong determined New Brunswick's highest elevation (820 metres) and named it Mount Carleton for Thomas Carleton (1736–1817), the first lieutenant-governor of New Brunswick. In 1969, a warden from a private fishing lodge drove me to the base (465 metres) of the mountain, and told me it would take 2½ hours to walk to the top. After 1¾ hours, I had reached its summit, and spent another hour viewing the grand vistas. Later, I asked the warden why he had not told me I could easily reach the top in less than two hours. In his reply, he revealed his respect for the mountain and his perceptive judgment of my nature: 'I wanted you to enjoy the climb without worrying how far it was to the top.'

In 1990, Clifford and Ruth Holtz, two Oshawa residents then in their sixties, set out to climb the highest elevations in each province and territory. They accomplished the ascent of all eleven the following year.

How Our National Parks Got Their Names

The names of Canada's national parks reflect a variety of origins. Some derive from a dominant local community (Banff, Revelstoke) or a physical feature (Riding Mountain, Gros Morne). Others come from a leading characteristic within the park's area (Glacier) or purpose (Wood

Buffalo), while some convey a general description (Pacific Rim, Cape Breton Highlands).

Canada's Aboriginal languages have contributed a number of park names (Kouchibouguac, Yoho, Kluane), with all of the recent parks in the north having either Inuktitut or Inuvialuktun names specially created for them (Auyuittuq, Tuktut Nogait, Ivvavik, Aulavik, Sirmilik, and Quttinirpaaq).

Rocky Mountains Park was the name given 100 years ago to the first reserve set aside in Alberta for the education and enjoyment of the nation. Subsequently the park area was enlarged, until 1930, when it was renamed Banff, an identification it had had informally for several years. Banff is the name of a Scottish town and county, and was given in either 1883 or 1884 to a CPR station a little east of the present townsite, established in 1886 by George Stewart, the park's first superintendent. Although Banff is closely linked to Lord Strathcona and Lord Mount Stephen, it is not certain if either influenced its choice as a station name; the latter was born in Banffshire.

Jasper, created in 1907, took its name from Jasper House, a North West Company trading post founded in the early 1800s. The post had been named for Jasper Hawes, who ran the post in 1817.

The name Yoho, given to a river, a mountain, a station, and a pass, as well as the park, is usually equated with the Cree for 'wonder' or 'excitement.'

The name Kootenay – the park embraces the headwaters of the Kootenay River – may have been given by the Blackfoot to describe the 'people of the water' in the language of the Kootenay. The 'ay' spelling was fixed in 1864 by Frederick Seymour, first governor of British Columbia, to differentiate it from Kootenai, the form used in the United States.

Mount Revelstoke – the park encloses some spectacular topography of the Selkirk Mountains – was named in 1914 for the city of the same name; the latter had been named in 1886 for Lord Revelstoke, head of a British banking house that had provided some of the financing for CPR construction. Glacier, also in the Selkirks, was created in 1886 and contains more than 100 glaciers.

Gwaii Haanas park reserve at the south end of Moresby Island (in the Queen Charlotte Islands) was set up as South Moresby park reserve in 1988, and its present name was confirmed in 1993. In Haida the name means 'islands of wonder and beauty.'

Waterton Lakes, on the Alberta–Montana border, dates from 1895. The lakes were named in 1858 by Thomas Blakiston for a noted British naturalist, Charles Waterton. The three lakes are joined by narrows called Dardanelles and Bosporus.

Wood Buffalo is Canada's largest park. It was established in 1922 to protect a small herd of wood buffalo and to provide a reserve for a larger herd of plains buffalo transferred from Wainwright, Alta.

Saskatchewan's Prince Albert National Park is north of the city of Prince Albert, which was named for the consort of Queen Victoria. Grasslands in southern Saskatchewan represents one of Canada's major ecosystems.

Riding Mountain, which occupies the rugged landscape west of Lake Manitoba, was created in 1930 and takes its name from the park's highest elevation. Riding Mountain probably derives its name from Aboriginal pack trails used by the early explorers.

Wapusk was established in 1996 on the shore of Hudson Bay, southeast of Churchill, to protect the dens of polar bears and other wildlife. The name in Cree means 'white bear.'

The names of two parks in southern Ontario (St. Lawrence Islands, Georgian Bay Islands) merely indicate their locations. Pelee is named for a prominent point extending into Lake Erie; the French word *pelée* means 'bald' or 'denuded,' a description given to the point in the early 1700s.

Pukaskwa is located on the north shore of Lake Superior in the area of the Pukaskwa River. The name's origin is uncertain, with 'fish cleaning place' being the most plausible. There is also a story that the name recalls an Aboriginal's killing and burning of his wife. The name is pronounced 'PUK-a-saw.' Bruce Peninsula park and the adjoining Fathom Five Marine park were created in 1987.

Québec has three national parks. La Mauricie is on the west side of Rivière St-Maurice, a name given in the eighteenth century for the Sieur de la Fontaine, Maurice Poulin. The river had been named 'Rivière de Fouez' in 1535 by Jacques Cartier for a French family named Foix; it was also known as 'Rivière des Trois-Rivières.' Forillon embraces the most easterly part of the Péninsule de la Gaspésie. The name is said to relate to the practice of fishermen using fires to attract fish or to alert ships to dangers in navigation.

The latest Québec park, Mingan Archipelago, consists of a cluster of small islands on the north shore of the St. Lawrence, 160 kilometres east of Sept-Îles. The name Mingan was long thought to be the Montagnais word for 'wolf'; more recently, the preferred derivation is that it comes from a Breton word to describe the islands' topography, 'rounded stone.'

New Brunswick has two national parks. Fundy is on the north shore of the Bay of Fundy. The name is likely a modification of 'fendu,' from the French name for Cape Split at the entrance to Minas Basin. Kouchibouguac, near Richibucto, was created in 1971. Taken from the Kouchibouguac River, the name is derived from the Mi'kmaq expression *Pijeboogwek*, 'river of the long tideway.'

Cape Breton Highlands, the older of Nova Scotia's two

national parks, takes its name from the most northerly part of Cape Breton Island. Cape Breton may have been named in the early 1500s for the fishermen from France's Bretagne. Kejimkujik was created in 1964. Its largest lake, Kejimkujik Lake, has a Mi'kmaq name of uncertain origin. Nova Scotians pronounce it 'KEJ-im-KOO-jik,' 'KEJ-ma-KOOJ,' and 'KEJ-a-ma-KOO-jee.'

Terra Nova in Newfoundland takes its name from a river and a lake, both derived from the Latin for 'new land.' Gros Morne is north of Corner Brook in the Long Range Mountains. Gros Morne, from the French for 'great height,' is the highest point in the park, rising to 806 metres.

Kluane in the Yukon is on the east flank of the spectacular St. Elias Mountains, which are interspersed with impressive icefields and glaciers. The name, pronounced 'KLOO-aw-nee,' means 'whitefish place' in Tlingit. Ivvavik, in the northern part of the territory, was called Northern Yukon in 1984, but was renamed ten years later after the Inuvialuktun word for 'place of giving birth and raising young,' in reference to the Porcupine herd of the barren-ground caribou.

Nahanni is a corridor along 300 kilometres of the South Nahanni River, a wild and turbulent tributary of the Liard River in the Northwest Territories. Nahanni is named for an Athapaskan tribe and means 'people of the west.' The newest parks in the Northwest Territories are Aulavik (1992) and Tuktut Nogait (1996). The former, on Banks Island, means 'where people travel' in Inuvialuktun and the latter, on the south shore of Amundsen Gulf and east of Inuvik, means 'calving ground of the young caribou' in the same language spoken by the local Inuit.

Auyuittuq was established as a park on Baffin Island in 1972. Its Inuit name means 'place that does not melt' in reference to its most prominent feature, the Penny Ice

Cap. Two national parks have recently been created in the northern lands of Nunavut: Quttinirpaaq (1999), formerly called Ellesmere Island park reserve, and Sirmilik (1999) on northeastern Baffin Island. The former's name in Inuktitut means 'top of the world,' and the latter's means 'the place of glaciers.'

A Medley of Favourite Names

Do you remember this sonorous phrase of the 1950s and 1960s: 'Jack Pickersgill, the honourable member for Bonavista–Twillingate'? The two place names spoken together are quite melodious, and conjure up pleasant images.

The town of Bonavista is located at Cape Bonavista, 130 kilometres northwest of St. John's, Nfld. The name of the cape is one of Canada's oldest, possibly having been given in 1500 by the Portuguese explorer Gaspar Corte-Real for Boa Vista, one of the islands of Cape Verde off Africa's west coast. The town of Twillingate, situated on South Twillingate Island in Notre Dame Bay, is 140 kilometres northwest of Bonavista. It was named by French fishermen for Pointe de Toulinguet, at the entrance to the harbour of Brest, in Bretagne, France.

Bonavista is now part of the riding of Bonavista–Trinity–Conception. Gander–Twillingate was the name of a riding until 1988, when it was changed to the more mundane Gander–Grand Falls.

These pleasant, musical names have inspired me to select my two favourite place names from each province and territory.

Two names in Nova Scotia that I particularly like are Antigonish and Pubnico. Antigonish, the home of St.

Francis Xavier University, takes its name from the Mi'kmaq word *Nalegitkoonech.* Its meaning may refer to how Rights River flows through broken marsh into Antigonish Harbour. The usual translation, 'where branches are torn off,' probably refers to the site of the town, and to an entirely different Mi'kmaq word.

Pubnico, 35 kilometres southeast of Yarmouth, is at the head of Pubnico Harbour. The 15-kilometre-long harbour has several communities on each side having Pubnico as part of their names. Among these are Lower West Pubnico, Middle West Pubnico, West Pubnico, Upper West Pubnico, Lower East Pubnico, Centre East Pubnico, Middle East Pubnico, and East Pubnico. The name comes from the Mi'kmaq *Pogomkook,* meaning 'dry sandy place.'

Rustico is my favourite in Prince Edward Island. It forms part of the names of several farming and fishing communities on the shores of Rustico Bay. Named for René Rassicot, who came from France's Normandie in 1724, Rustico became the local form by 1774.

I also like Strathgartney, the name of a provincial park 17 kilometres west of Charlottetown. In 1846, Robert Bruce Stewart selected the name for his homestead, naming it after a valley in Perthshire, Scotland.

Among my favourites in New Brunswick is Nackawic, a village at the mouth of Nackawic Stream. The name is derived from the Maliseet *Nelgwaweegek,* suggesting 'straight water,' because of the way the stream entered the Saint John River before the river was dammed.

Kouchibouguac is a little village on the north bank of the Kouchibouguac River, 40 kilometres southeast of the city of Miramichi. In 1971, Kouchibouguac National Park was created along the seashore from Kouchibouguac Bay to Richibucto Harbour. The name means 'river of the long tideway' in Mi'kmaq.

Sorel, where the Richelieu flows into the St. Lawrence

northeast of Montréal, is my favourite Québec place name. The site was granted in 1672 to Pierre de Saurel, a captain of the Carignan-Salières Regiment. Called Sorel before 1787, it was renamed William Henry for Prince William Henry, future King William IV, who visited the place that year. The name Sorel was restored in 1860, and by 1906 archivist Pierre-Georges Roy reported that William Henry was totally forgotten.

My other Québec choice is the town of Maniwaki, located on the Gatineau, 125 kilometres north of Ottawa. The name, derived from the Algonquin for 'Mary's land,' was given in 1849 by the Oblates of Mary Immaculate.

Ontario's Muskoka evokes images of summer cottages, boat trips among the islands, and rambles in the woods. Lake Muskoka got its name from two Ojibwa chiefs, Yellowhead and his son William Yellowhead, both of whom were known as Musquakie in their language. Both Musquakie and Muskoka were derived from *mesqua ahkees*, 'red ground.'

I like both the name Cobalt and the indomitable spirit of the people of this Northern Ontario town. Silver was discovered there in 1903. Dr. Willet Miller, the provincial geologist, went to the site in 1904 and chose the new railway station's name from traces of the mineral cobalt found near the station.

During my childhood, I read the name Wawanesa on a calendar in my grandfather's house and imagined the home of the national insurance company would be a great city on the Prairies. In fact, the village, 35 kilometres southeast of Brandon, Man., has never been the metropolis I envisioned, and now has fewer than 500 people. The name probably means 'whippoorwill' in Algonquin; the bird is mentioned in Longfellow's poem *The Song of Hiawatha* as *Wawonaissa.*

Minnedosa, a town 45 kilometres north of Brandon, is

a mellifluous name created in the 1880s by a miller named J.S. Armitage. He derived it from the Siouan words for 'swift water,' in reference to the fast-flowing Little Saskatchewan River. The river, which flows through the town to the Assiniboine River, was officially called Minnedosa River in 1928, but the preferred local name, Little Saskatchewan River, was restored in 1980.

Assiniboia, 90 kilometres south of Moose Jaw, is one of my favourite Saskatchewan names. Formerly called Leeville for the first postmaster, Hubbard Lee, the town was renamed in 1913 for the former District of Assiniboia, part of the North-West Territories from 1882 to 1905. The district was named for the Assiniboine River, itself given for the Siouan tribe whose name in Ojibwa means 'one who cooks with (hot) stones.'

The centre of Saskatchewan's potash industry is Esterhazy, a town of 3,000 south of Yorkton. The name honours Count Paul Esterhazy, a Hungarian who, while living in New York in the 1880s, resolved to help his fellow countrymen working in desperate conditions in the Pennsylvania coal mines. He bought land north of the Qu'Appelle Valley and persuaded many of the miners to move there. Later they were joined by farm families who came to Canada from Hungary.

Pincher Creek is one of my favourite names in Alberta because it reflects the spirit of the wild west. The town of Pincher Creek, with a population of 3,650, is located on a tributary of the Oldman River, 85 kilometres west of Lethbridge. Its name recalls the loss of a pair of pinchers (an older form of 'pincers') in the 1860s by a prospecting party. Another story is that some of the prospectors were murdered by Aboriginals. When the remaining members of the party came across the grisly site, they found the pinchers of one of their murdered friends near the creek.

Between Edmonton and Red Deer is the attractive

town of Lacombe, with a population of 5,600. The name recalls the missionary Albert Lacombe (1827–1916), who ministered to the Aboriginals and the Métis from Winnipeg to Lesser Slave Lake during the last half of the nineteenth century. In the 1880s, Father Lacombe, who was partly of Aboriginal origin, helped the CPR reach an amicable settlement with Chief Crowfoot for the construction of the railway across Blackfoot lands in southern Alberta.

One of my favourite British Columbia names is Kitwanga, a village in the Skeena River valley between Terrace and Hazelton. I find its lilting syllables, 'KIT-wawn-GAH,' especially pleasant. The name is derived from a Gitksan phrase meaning 'people of the place of rabbits.'

Keremeos, a village of 700 about 40 kilometres southwest of Penticton, is another name I find appealing to the ear. It was derived in 1887 from an Okanagan word, possibly meaning land 'cut across in the middle' or 'a flat cut through by water' in reference to the Similkameen River crossing the flat lands near the village.

No name sings more about the Yukon and the Gold Rush days than Klondike. The name, from *Thron-duick*, meaning 'hammer water' in an Athapaskan dialect, relates to the practice of putting stakes in the river to catch salmon. Ironically, the fishing was so poor in the summer of 1896 that George Carmack and his party abandoned their stakes to prospect in one of the Klondike's tributaries, the legendary Bonanza Creek, and the stampede was on.

Whitehorse has a certain beauty and strength in its name. It is said that the first miners travelling down the Yukon River in 1880–1 thought the rapids at the site of the present city looked like the manes of white horses. The builders of the White Pass and Yukon Route first called the townsite Closeleigh in 1899, but reverted in

1900 to the name that had already become widely known throughout the world. In the early 1950s, Whitehorse replaced Dawson City as the territorial capital.

Another territorial capital, 1,000 kilometres to the east, has another of my favourite names. Gold was discovered on the north shore of Great Slave Lake in 1934. The community that sprang up at the site was named Yellowknife, derived from the name of a band of Athapaskans who made tools from yellow copper. Yellowknife, with a population of almost 17,000, has been the capital of the Northwest Territories since 1967.

Nanisivik is a small mining community on Baffin Island, in Nunavut. It was developed in 1974 by Nanisivik Mines to extract lead, zinc, silver, and cadmium. In Inuktitut, the name means 'place where people find things.'

So there you have my twenty-four favourites. Half are names of Native origin, revealing my preference for names derived from the languages of the First Nations of Canada. There are no -ville, -town, -burg, or saint names among my selections. Although some are names transferred from other parts of the world, none is derived from places in England or the United States.

Riding Names: Same Gratifying, Some Grating

In the essay 'A Medley of Favourite Names,' I wrote that Bonavista–Twillingate used to be a favourite name of mine for a federal electoral district, and that Twillingate was now preserved in the name of Gander–Twillingate. Wrong. This was a riding name prior to the redistribution that came into effect in July 1988. The new riding was called Gander–Grand Falls.

The 1988 redistribution provided for 295 federal ridings, up from 282 in the previous Parliament. Of these, 107 were new riding names, while 188 were the names of the redistribution as they existed in July 1988. The next redistribution, based on the 1991 census, resulted in 301 ridings, with 6 new names and a whopping 115 changes.

The creation of boundaries and the selection of names for each electoral district are the responsibility of commissions appointed in each province and territory following each decennial census. When a new district corresponds generally to a riding at the time the reports are tabled in the House of Commons, a sitting member's riding name may be changed or restored by a private member's bill. Following an election, a change in a riding name can also be made through a private member's bill.

After redistribution based on the 1971 census, more than forty name changes were accepted by the House of Commons. After the 1988 redistribution, there were eighteen changes to the new riding names to June 1993. The commission in Québec had recommended Chapleau for the area of the former district of Gatineau, but the sitting member successfully appealed in the summer of 1988 to have it renamed Gatineau–La Lièvre. In 1996, the single name Gatineau was restored. In New Brunswick, the commission suggested Chaleur in place of Gloucester, in order to reflect the francophone majority and to commemorate the name given by Jacques Cartier in 1534. However, Gloucester, in continuous use since 1867, was retained after an appeal by the sitting member. Subsequently, in June 1990, it was changed to Acadie–Bathurst. The riding of Restigouche was changed to Restigouche–Chaleur in 1989, and became Madawaska–Restigouche in 1996.

Among the most attractive riding names are single

place names that clearly identify the specific area of the country. Examples are Athabasca, Peterborough, Skeena, Wetaskiwin, Chicoutimi, Thornhill, and Yukon. Also pleasing are riding names comprising two names that produce a musical cadence. Among these are Brome–Missisquoi, Hochelaga–Maisonneuve, Cariboo–Chilcotin, Regina–Qu'Appelle, Winnipeg–Transcona, Edmonton–Strathcona, Fundy–Royal, and Timmins–Chapleau. But as more place names are added to a riding name, it becomes rather cumbersome, as in the case of Dufferin–Peel–Wellington–Grey and Beauport–Montmorency–Côte-de-Beauport–Île-d'Orléans.

St. John's East and St. John's West have been in use since 1949, but as several communities in the former are *west* of the latter, new names such as St. John's–Torbay and Avalon–Placentia would have been better choices. In the redistribution based on the 1971 census, Humber–Port au Port–St. Barbe was an awkward title. Its successor is equally awkward: Humber–St. Barbe–Baie Verte.

Prince Edward Island has four ridings, each with a single name chosen from a prominent physical feature. Although the single name has its appeal, each is rather vague as to the geographical area to which it refers. For example, Charlottetown would be more explicit than the current name of Hillsborough.

Six riding names in Nova Scotia were changed in 1996. Several of the province's eleven riding names were either vague or inappropriate in 1988. Central Nova and South West Nova were rather bland. Two names were quite misleading: the Cape Breton Highlands, were for the most part, not in the riding of Cape Breton Highlands–Canso, and the main section of the Annapolis Valley was not in Annapolis Valley–Hants. The Nova Scotia commissioners resolved the discrepancies in 1996 by introducing Sydney–Victoria and Bras d'Or–Cape Breton on

Cape Breton Island, and renaming Annapolis Valley–Hants as Kings–Hants.

In New Brunswick, Fredericton was substituted for York–Sunbury, Miramichi for Northumberland–Miramichi, and Beauséjour for Westmorland–Kent in 1988. The province's commission believed that parts of the previous names could be confused with riding names in Ontario. Beauséjour, derived from the name of a French fort built in 1750, destroyed by the British in 1755, and restored as a national historic park in 1926, was a pleasant name for this riding which takes in a large area to the north and east of Moncton, including the university town of Sackville on the Isthmus of Chignecto. In 1996, the riding name was changed to Beauséjour–Petitcodiac.

The commission in Québec has shown sensitivity for traditional names, with twenty-two of the original sixty-five names of 1867 still in use. Among the more pleasing to the ear are Chambly, Charlevoix, Portneuf, Charlesbourg, Shefford, and Lotbinière.

The commission tried in 1988 to restore the boundary of Québec–Est to include the heart of the old city, with Québec–Ouest to the west of it. But opposition persuaded the commission to maintain Langelier, in use only since 1966, in the downtown area, and to continue the use of Québec–Est to the *west* of the city centre. In 1990, Langelier became Québec (French) and Quebec (English). For the two, Québec–Limoilou and L'Ancienne-Lorette–Vanier would have been suitable alternatives. Lachine–Pointe-Claire might be preferable to Lac-Saint-Louis. Sainte-Foy–Sillery would be more explicit than Louis-Hébert. As Montréal-Nord is the only municipality comprising Bourassa, the municipal name might be a better choice.

Ontario has two unwieldy riding names. Hastings–Frontenac–Lennox and Addington, formed from parts of these three counties, is quite a mouthful. For it, Napanee–Bancroft might be a suitable alternative. For the ungainly Dufferin–Peel–Wellington–Grey, an innovative choice might be Ontario Highlands, the word *highlands* being widely used to describe this highest part of southern Ontario.

Simcoe North could well be named Huronia, which has strong historical associations with the Midland and Orillia areas, and is used as a regional name in weather reports. Quinte would be an apt substitute for Prince Edward–Hastings. The area of the new riding of Ottawa South would be understood better if the name Ottawa–Alta Vista were used, since the neighbourhood traditionally known as Ottawa South is actually in the riding of Ottawa Centre.

Brandon–Souris and Dauphin–Swan River are two of my favourite riding names in Manitoba. Unfortunately, five of the province's electoral districts have directional names, with two of them, Winnipeg North Centre and Winnipeg South Centre, being unwieldy and unimaginative. Four of these names date back to 1924, but the choice of Assiniboine, Fort Garry, or Kildonan for these ridings would better reflect the rich history of the Red River Valley.

The fourteen electoral districts of Saskatchewan were completely reorganized in 1988, with only three names of the previous redistribution being kept. In 1996, only four of those names continued unchanged. The commissioners avoided directional words, choosing instead such excellent names as Regina–Qu'Appelle, Cypress Hills–Grasslands, Yorkton–Melville, and Saskatoon–Humboldt. The commissioners gave no reason in their report for selecting Saskatoon–Clark's Crossing. It was changed to Saskatoon–Rosetown–Biggar in 1996.

There are some colourful names in Alberta outside the

two main cities. Crowfoot, from the name of the legendary Blackfoot chief, and Wild Rose, from the provincial flower, are imaginative, although neither is especially identified with its respective area. Wild Rose, in the rural area north of Calgary, was first named Kneehill in 1988 by the Alberta commission, but when it was pointed out that this is the name of a district municipality mostly in the adjacent riding of Crowfoot, the commissioners chose Wild Rose instead.

Eleven of Alberta's twenty-six electoral divisions have directional components in their names. The commission claimed that geographical directions are more easily recognized by the public, but Calgary–Glenmore (from Glenmore Reservoir and Glenmore Trail) might be just as effective as Calgary Southwest, and Calgary–Blackfoot (from Blackfoot Trail) as well understood as Calgary Southeast. It is pleasing that Edmonton–Strathcona was kept. Perhaps better alternatives could be found for Edmonton East, Edmonton Southeast, and Edmonton Southwest.

British Columbia has some of the most evocative riding names, such as Vancouver–Quadra, Nanaimo–Cowichan, Cariboo–Chilcotin, and Esquimalt–Juan de Fuca. In 1996 Okanagan Centre was renamed Kelowna, and Okanagan–Similkameen–Merritt became Okanagan–Coquihalla. Fraser Valley East became simply Fraser Valley, while Fraser Valley West was renamed Langley–Abbotsford.

In 1988, one riding name in Québec was given both English and French forms: Mount Royal and Mont-Royal, which are statutory forms of the city's name. In 1990, four more Québec ridings were assigned bilingual names. Otherwise, forms like Témiscamingue, Trois-Rivières, and Bonaventure–Gaspé–Îles-de-la-Madeleine–Pabok are respected in English. Elsewhere almost all the official French names of ridings with directional words are translated, such as Edmonton–Sud-Ouest, York–Sud-Weston, Surrey-Est. Some names are awkward hybrids, such as Ouest Nova, St. John's–Ouest, and Fraser Valley–Ouest. Adding accents to some names (e.g., Nanaimo, Vegreville, Malpeque) goes against the official forms of these names. Saanich–Gulf Islands had been translated to Saanich–Les Îles-du-Golfe in 1988, but in 1996 its English form was retained as its French title. Mysteriously, names with river (e.g., Peace River and Rouge River) have been saved from translation.

Twillingate may be gone from the name of a federal riding, but we still have such distinctively Canadian names as Skeena, Medicine Hat, Nunavut, Nickel Belt, Hochelaga–Maisonneuve, and Bonavista–Trinity–Conception.

'Hogtown' and Other Monikers across Canada

Nicknames of places are common across the land. Some are boastful, others are honorific, and a few are downright nasty.

- Promotional slogans abound, like 'Ambitious City' for Hamilton; the 'Honeymoon Capital of the World' for Niagara Falls; 'Hub of the Maritimes' for Moncton; 'Gateway to the North' for each of Edmonton, Prince Albert, and North Bay, Ont.
- Affectionate labels often adorn places, such as 'City of Bridges' for Saskatoon, 'Celestial City' for Fredericton, 'Royal City' for both New Westminster and Guelph, 'Cradle of Confederation' for Charlottetown, 'Loyalist City' for Saint John, N.B., 'Queen City' for both Regina and Toronto, and 'Forest City' for London, Ont.

- Glorified comparisons are often made with foreign places, such as referring to Montréal as the 'Paris of Canada,' Winnipeg as the 'Chicago of the Canadian West,' Ottawa as the 'Washington of the North,' Hamilton and Sydney, N.S., as the 'Pittsburgh of Canada,' Biggar, Sask., as the 'Little Apple' (the town has been noted on signs since the 1940s with 'New York is big, but this is Biggar,' leading to another comparison with America's 'Big Apple'), and Toronto as both the 'Belfast of America' and the 'Athens of the Dominion.'
- Some names are clipped to a single syllable, examples being 'The Hat' for Medicine Hat, 'The Peg' for Winnipeg, and 'The Soo' for Sault Ste. Marie.
- Significant characteristics of a place can result in names like 'Steeltown' for Hamilton, 'Limestone City' for Kingston, 'Flowertown' for Brampton, and 'Motor City' for both Oshawa and Windsor.
- Previous names for places are sometimes used to identify either an admirable history or a backward characteristic, such as 'Pile o' Bones' for Regina, 'Gastown' for Vancouver, 'Bytown' for Ottawa, and 'Muddy York' and 'Little York' for Toronto.
- Some places are often referred by their acrostic (the first letter of each word or fused words), examples being 'P.A.' for Prince Albert, 'P.G.' for Prince George, 'Y.K.' for Yellowknife, and 'T.O.' for Toronto (which may be an acrostic for Toronto, Ontario, or a false acrostic from the two initial letters of the name).
- A province or part of a province can have a nickname, such as southwestern British Columbia being called 'Lotus Land,' and the Island of Newfoundland having become well known as 'The Rock.' Prince Edward Island is known as 'Garden of the Gulf,' 'Million Acre Farm,' and 'Spud Island.'
- Derisive titles are sometimes attached to places by out-

siders, examples being 'Cowtown' for Calgary, 'Fat City' for Ottawa, and 'Winterpeg' for Winnipeg.

No place in Canada has as many sobriquets as Toronto. As well as those noted above, the city has often been called the 'City of Churches' and 'Toronto the Good,' each of which was not necessarily complimentary. Envy of places south of the border has spawned 'Broadway North' and 'Hollywood North.' In 1975 *Washington Post* correspondent Anthony Astrachan called it 'City that Works' in an article in *Harper's*. Recently it has been referred to as the 'Big Smoke,' which may have originated in an Australian Aboriginal comment about the big cities Down Under, and was picked up by *Maclean's* writer Allan Fotheringham, who may have thought the phrase appropriate for a place with 'big reputation, little to show for it.' Toronto is fortunate in that references, especially the negative ones, to the government and legislature of the province of Ontario are usually made to 'Queen's Park,' and not the designated provincial capital. Sometimes Toronto has been dubbed the 'Centre of the Universe.'

At the end of the nineteenth century, the city of Cincinnati had acquired the nickname 'Porkopolis' because it was a leading hog-slaughtering city in the United States. Its equivalent in Canada was Toronto, where the British Empire's finest bacon and pork earned the city the sobriquet 'Hogtown.' Or so the story goes, even though there appears to be scant evidence that the nickname was ever used to describe Toronto before the 1930s. The earliest citation that I have found is on a card dated 17 May 1943 in the Metropolitan Toronto Reference Library: 'Why is Toronto called Hogtown? Mr. T.A. Reed (1871–1958), authority on Toronto, says that in his opinion the name had no definite origin, but was probably one applied in contempt as to a rich, selfish community which "hogged" things and the name stuck.' *A Dictionary of Canadianisms on Historical Principles* (1967) provides a 1959 reference from Vancouver's monthly *Press* as its earliest citation for 'Hogtown.'

Without providing any citations, some have suggested that the moniker may be traced back to 1849 when the Toronto city council passed a bylaw prohibiting hogs from running at large within the municipal limits. Others have suggested that 'Hogtown' evolved in the late nineteenth century when Joseph (later Sir Joseph) Flavelle made the William Davies Company a leading pork packer. University of Toronto historian Michael Bliss wrote a comprehensive biography of Flavelle titled *A Canadian Millionaire* (Toronto: Macmillan, 1978), and in it entitled a whole chapter 'Hogtown,' concluding that in the early years of the twentieth century 'people were starting to call Toronto "Hogtown."' Unfortunately, Bliss's otherwise excellent scholarly biography provides nary any solid evidence that the moniker was ever used then. Professor J.M.S. Careless, writing in 1984, mentioned how the Davies's 'bacon empire may well have inspired Toronto's later nickname "Hogtown."' Geographer Michael Doucet of Ryerson Polytechnic University has extensively studied the nicknames given to Toronto, and has concluded that 'Hogtown' emerged when Toronto became a leading meat-packing centre along its waterfront, later in the city's west end, and then was reinforced by the media after about 1945 in other Canadian cities, where the word 'Hogtown' has been used to identify Canada's greediest and snootiest urban centre.

In 1995 the *Toronto Star*'s 'Words' columnist Lew Gloin devoted four columns to the origin of 'Hogtown.' He wondered if the nickname had come about because hogs were observed rooting near muddy Yonge Street's waterfront in the early 1800s. He referred to the recollection of Dick Griffin of Norland (north of Lindsay) that 'Hoggtown' (Griffin's spelling) came about because a man named Hogg farmed at Hogg's Hollow just beyond the northern limits of the city, but again without any written evidence. Gloin concluded that majority opinion favoured meat-packing as the ultimate source of 'Hogtown.'

Some writers believe that 'Hogtown' has been fading from use over the past forty years. That was Pierre Berton's conclusion in *The New City* (1961), Doug Gloin's in the *Toronto Star* in June 1995, and Robert Sheppard's in the *Globe and Mail* in April 1997. In observing the mercurial change taking place in Toronto, Berton said that it was 'possible to look back on Hogtown with warmth and affection.' Gloin claimed that 'Toronto's old nickname, Hogtown, has all but disappeared, surfacing only occasionally in headlines on the sports pages of local papers.' Sheppard wrote: 'Welcome to the new city that used to be called Hogtown and that is now so much more like a pig in a poke.' It may be true that the nickname has declined in use within the present city, but it still gets mentioned often in the press across Canada to refer to the perceived selfishness and smugness of the city on the north shore of Lake Ontario.

I get the sense that Canada's only truly cosmopolitan city, the only one that can comfortably wrap itself in its own adulation as being widely respected by other world cities of a similar size, and does not in any way resent being called 'Hogtown' – it is a fact of its history. As Bruce West wrote in 1967, Torontonians take a peculiar pride in the way people talk about them.

Scrutinizing Political Issues and Language Problems

Of Hatters and Capers, Townies and Trifluviens, and Other Monikers People Call Themselves

Haligonian, Montrealer, Torontonian, Winnipegger, Calgarian, Vancouverite. These are abbreviated ways of referring to the residents of places, but there does not appear to be a distinct term in English to describe such words. Some dictionaries offer *gentilitial* and *ethnonym*; they are rarely used. Some writers have used *ethnic* and *dominym*, but these are not very helpful in explaining the concept of people living in particular locations. I have suggested *patrial*, without success, as well as *nopitacs* (names of people in towns and cities), but these have not been received with much enthusiasm either.

Francophones use the word *gentilé* to describe such words as Parisien, Montréalais, and Gaspésien. The Commission de toponymie du Québec has ruled on the correct *gentilés* for most municipalities in the province; these terms are listed in their masculine and feminine forms in the government's municipal directory, *Répertoire des municipalités du Québec.*

Since the English language does not appear to have a specific word to describe the residents of a geographically defined place, I will use the word *gentilé* (pronounced 'JAWN-tee-LAY'). All the provinces and the Yukon have distinctive *gentilés* in English. Some, such as Newfoundlander and Albertan, possess a very strong sense of historical and cultural identity. The Northwest Territories does not lend itself to a *gentilé*, and Nunavut is still too young to have acquired one, although Nunavutan may be an obvious choice.

The *gentilé* for a person from the province of Québec – Québécois – is spelled both Quebecer and Quebecker in English. The term Québécois in English usually implies a Quebecer whose language and culture are distinctively French Canadian.

In Québec, a few *gentilés* have been endorsed in English, such as Shawvillite, but most follow the rules of the French language; for example, Ayer's-Cliffois. Several are derived from Latin roots, such as Trifluvien (Trois-Rivières) and Fidéen (Sainte-Foy). A few reflect the place's history, with Havre-Saint-Pierre having Cayen from its Acadian roots, and Sainte-Élisabeth having Bayollais, from the full name of St. Elizabeth of Bayolle.

There are six different ways to form a *gentilé* in English; the final syllable of the source name usually provides a clue as to whether it should take *n, ian, onian, ite, er,* or *ese,* the last apparent in Canada. Names ending in *a* and *ia* add an *n*: thus we have Oshawan, Mississaugan, Kelownan, and Victorian. An exception is a resident of Kenora, dubbed a Kenoraite by Mark Orkin in *Speaking Canadian English,* published in 1970, and confirmed in 1989 by the editor of the *Daily Miner and News.*

The *onian* suffix is almost always used by places ending in *ton,* resulting in Monctonian, Frederictonian, Edmontonian, and Hamiltonian. A number of other places, with no apparent relationship in their final syllable, also have the *onian* ending, such as Haligonian (Halifax),

Amherstonian, Galtonian, Woodstonian, Saskatonian, and Pictonian, the last for both Pictou, N.S., and Picton, Ont. While Orkin gives Bellevonian for Belleville, the editor of the city's newspaper, *The Intelligencer,* stated in 1989 that he prefers the term Bellevillian.

In the United States, names ending in *town* usually form their *gentilé* by using *tonian,* but the preference in Canada may be to add *er,* as in Charlottetowner.

Names ending in consonants and silent vowels form their *gentilé* in one of three ways: by adding *ian, ite,* or *er.* Usually names with *ville* drop the *e* and add *ian,* but sometimes such names also end in *ite,* especially if residents suspect that an outsider may invert the *a* and the *i* and refer to someone from Brockville as a 'Brockvillain.'

At one time, names with *ian* were likely more numerous in Canada than those with *ite,* but *gentilés* with the latter seem to be growing in popularity. Vancouver residents were once known as Vancouverians, but Vancouverite is now universally used. Among Canadian *gentilés* with the *ian* ending are Stratfordian, Fort Erian, Bathurstian, Yarmouthian, Sudburian, Kamloopsian, and Calgarian. People in Peterborough are called Peterburians. The residents of the capital of Newfoundland have been referred to as St. Johnsians, but they are really better known as 'townies,' as opposed to 'baymen,' who live in outport communities.

Names with *ite* include Sydneyite, Granbyite, Cornwallite, Guelphite, St. Catharinite, Brandonite, Thompsonite, and Banffite, but these forms are rarely used. Orkin adds Barrie-ite, Sooite, and Nanaimoite, but they are not common.

Among names with the suffix *er* are: Saint Johner, Lunenburger, Summersider, Londoner, Winnipegger, Flin Floner, Medicine Hatter (often simply Hatter), and Yellowknifer.

Some places may not have a *gentilé* at all, especially those with two or more words, like Niagara Falls, North Bay, and Thunder Bay, although some people have told me that Niagara Fallsian, North Bayite, and Thunder Bayite have been used by the media. Residents of Thunder Bay may be more familiar with Lakeheader.

Sometimes the residents of several adjacent communities use a common *gentilé.* The people living on the island of Montréal between Dorval and Sainte-Anne-de-Bellevue are collectively called West Islanders. Residents from Port Hawkesbury to Glace Bay, and beyond to Ingonish and Inverness, Nova Scotia, and all the villages and crossroad communities in the region, are proud to be called Cape Bretoners or Capers. Labradorian is a badge of honour for the long-time residents of Newfoundland living on the mainland side of the Strait of Belle Isle. The people of Prince Edward Island are known simply as Islanders, sometimes as Spud Islanders. A Gulf Islander is someone who has settled into the bucolic life on one of the islands in the Strait of Georgia between mainland British Columbia and Vancouver Island.

A *gentilé* that does not seem to fit any pattern is Moose Javian, although it may be related to Shavian, used in referring to the literature and ideas of George Bernard Shaw. I have also seen the form Moosichapishanisippian, given in jest by a former Reginan.

Many people are not above composing humorous *gentilés* for various places, among them being Niagara Felons, Fort Erings, Smithereens, and Terracites (for Smithers and Terrace, towns in northwestern British Columbia), and Bragg Crickets (for Bragg Creek, a community near Calgary).

There appears to be a large number of places, with populations exceeding 10,000, that do not possess a *gentilé* in English. They include Corner Brook, Nfld.; Truro,

N.S.; North Battleford, Sask.; Grande Prairie, Alta.; Prince Rupert, B.C.; and Whitehorse in the Yukon.

Pronouncing Names as the Locals Do

Queries are often received by names offices as to the correct pronunciation of certain place names. Other than suggesting the range of suitable pronunciation, names authorities invariably avoid making pronouncements on pronunciation.

Some names permit a range of pronunciation. While there may be a general preference for three pronounced syllables for Toronto ('t-RAHN-to'), to call it 'TRAWN-to' is not considered entirely wrong. Fredericton is called 'FRE-drick-tun' and 'FRE-dik-tun,' and Calgary is pronounced both as 'KAL-ga-ree' and as 'KAL-gree,' with the first in each instance appearing to be more correct and the latter somewhat sloppy diction.

Sometimes, the same name is pronounced differently in separate parts of Canada. Dalhousie University in Halifax and Dalhousie, N.B., are usually pronounced 'dal-HOW-zee,' but the street in Ottawa and the township in Lanark County, Ont., are heard as 'dal-HOO-zee,' and Port Dalhousie in St. Catharines comes out as 'port-da-LOO-zee.'

In Prince Edward Island it is 'SOO-ree' for the town of Souris, but the same name in Manitoba is 'SOO-ris.' In Nova Scotia, Greenwich is pronounced 'GREE-nich,' in New Brunswick it is 'GREEN-wich,' and in Prince Edward Island, 'GREN-ich.' The various Elgins in Canada are usually pronounced with a hard 'g,' but Elginburg near Kingston, possibly influenced by American pronunciation, is usually called 'el-JIN-berg.' St. John's, Nfld., is generally pronounced 'sint-JAHNZ,' but Saint

John, N.B., comes out with equal emphasis on both syllables: 'SAYNT-JAWN.'

Some names have had distinctly separate developments in terms of spelling and pronunciation, with the spelling form appearing to be unlike the pronounced form. Such a name is Magaguadavic River in New Brunswick, which was fixed in its spelling at the end of the eighteenth century, but has long been pronounced as 'ma-ga-DAY-vee.' The pronunciation of the nearby Digdeguash River ('DIG-gi-dee-WASH') is reflected in the spelling of another New Brunswick name: Diggity Stream.

There is the classic case of Bay d'Espoir in Newfoundland, which has long been ironically known as 'BAY-dis-PAYR.' Such a pronunciation, however, reflects seventeenth-century French speech. Etobicoke, the name of the former city on the west side of Toronto, but now part of Toronto, may come out of a stranger's lips as 'ee-TO-bee-KOKE,' but locally the form 'e-TO-bee-KO' is preferred. The name of the former rural city of Nanticoke (now part of the town of Haldimand), on Lake Erie near Simcoe, is pronounced 'NAN-te-KOKE.' Stouffville, near Toronto, is called 'STO-vul,' not 'STOOF-vil.' Gleichen in Alberta is pronounced 'GLEE-shun,' although the CPR financier for whom it was named, Count Albert Gleichen, pronounced his name 'GLAYK-kun.'

There are some names whose pronunciations seem to be designed to separate the local residents from outsiders. An example is Newfoundland, which is locally pronounced 'NYOO-fin-LAND,' not 'noo-FOWND-land.'

A pronunciation that certainly distinguishes a westerner from an easterner is the mouthing of Saskatchewan. A native barely enounces the first, third, and fourth syllables, saying 'sis-SKATCH-eh-win,' while an easterner is inclined to stretch it out to 'sas-SKA-choo-wahn.'

Another interesting example is North Gower, a community 16 kilometres south of Ottawa's city centre. Residents rhyme it with 'door,' not 'power.' Antigonish in Nova Scotia is 'an-ti-ga-NISH' not 'an-TIG-a-nish.' Maugerville east of Fredericton is called 'MAY-jer-vil' not 'MAW-ger-vil.' One can never be too certain about the letter 'g' in names. Margaree in Nova Scotia has a hard 'g' but Langenburg in Saskatchewan has a soft 'g' in the middle and a hard 'g' at the end.

Many names of French origin, or Native names that have come into English via French, have acquired some unusual pronunciations. Examples are Rencontre, Nfld. ('ROWN-kown-ter'); Pouce Coupe, B.C. ('poos-KOO-pee'/'poos-KOO-pay'); Quesnel, B.C. ('kwe-NEL'); Des Joachims, Que. ('de-SWI-shah'); Tobique River, N.B. ('TOH-bihk'); Bienfait, Sask. ('BEEN-FAYT'); Quirpon, Nfld. ('KAR-poon'); Grand Bruit, Nfld. ('gran-BRIT'); Punnichy, Sask. (poo-nih-SHY); Hobbema, Alta. (hoh-BEE-mah); Skidegate, B.C. (skih-dee-GET); Hecate Strait, B.C. (HEK-ut); Tête Jaune Cache, B.C. (tee-ZHAWN-kash); Cramahe Township, Ont. (KRAH-mee); and L'Ardoise, N.S. ('LORD-WAYZ').

In this sampling there are no doubt some perceived errors in the suggested pronunciations. As well, some unusual names have been omitted because of uncertainty about local preferences. In this group are Trepassey and Aquaforte, Nfld.; Gabarus and Necum Teuch, N.S.; Apohaqui, N.B.; Breadalbane, P.E.I.; Nicabong and Kazabazua, Qué.; Caesarea, Wanapitei, and Schreiber, Ont.; Melita, Man.; Wauchope, Sask.; Provost, Alta.; Ucluelet and Tsawwassen, B.C.; Rancheria, Y.T.; and Aklavik, N.W.T.

As with personal names, we all like to pronounce places as they are spoken by the local population. The stranger is usually forgiven, but announcers on radio and television are expected to know the local preferences.

Sorting Out the Sound-Alikes and Look-Alikes

Do you, like many of us, sometimes get confused by the look-alike, sound-alike cities of St. John's and Saint John? Do you know which one is the capital of Newfoundland? Do you wish they had more distinctive names so you would have no doubt which is which?

St. John's, the chief city and capital of the province of Newfoundland, traditionally takes its name from the discovery of its harbour by John Cabot on the feast of St. John the Baptist, 24 June 1497. The first reference to the name of the harbour, *R de Sam joham*, is on a map produced by Pedro and Jorge Reinel in 1519. The earliest mention of the city's name in its present form is in a letter written in 1527 by John Rut, who led an English expedition to North America that year. With a name having 460 years of continuous use, St. John's could hardly be expected to surrender it for another.

The Saint John River in New Brunswick was named *Rivière saincte-Jean* by Sieur de Monts and Samuel de Champlain on 24 June 1604. The city that developed at the river's mouth in one of Canada's most scenic settings – albeit often fogbound – was known as Menagoueche ('little peninsula') by the Mi'kmaq and the Maliseet, and also as Ouigoudi ('village') by the Maliseet. On the arrival of the Loyalists in 1783, two names came into use: on the east side of the river's mouth, Parr Town (for John Parr, governor of Nova Scotia) and, on the west side, Carleton (for Sir Guy Carleton, then commander-in-chief of British forces in North America, not for his brother, Thomas Carleton, first lieutenant-governor of New Brunswick). Two years later, Ward Chipman, a distinguished lawyer and politician, proposed the name Saint John for the first incorporated city in what is now Canada, and for the county extending along the Bay of Fundy shore.

In English, St. John's and Saint John receive some distinction in their spellings, the Newfoundland city having an apostrophe *s*, and the New Brunswick city always having its *saint* spelled in full. There is also a small distinction in their pronunciation: 'sint-JAHNZ' for the one; and, with equal emphasis on the two syllables, 'SAYNT JAWN' for the other. The unofficial French form, Saint-Jean, is often used for both, but it is not authorized for federal maps and documents.

Saint-Jean-sur-Richelieu, a city 40 kilometres southeast of Montréal, is the site of the former well-known Collège militaire de Saint-Jean. Once known in English as St. Johns, this form was officially dropped in 1962.

No doubt many people groaned in 1965 when Premier Joseph Smallwood of Newfoundland announced that Hamilton River in Labrador would be renamed Churchill River for Sir Winston Churchill, ignoring the fact that Canada already had another major river by the same name. The latter, which drains much of northern Saskatchewan and Manitoba, had been named in the late 1600s for John Churchill, Duke of Marlborough, a governor of the Hudson's Bay Company. An effort to have the name Hamilton River restored, while maintaining the widely known Churchill Falls, did not gain support.

About 1985, I read a report that a person from Toronto was going to Sackville, N.S., to give a lecture. My first reaction was that this was an error for the home of the well-known liberal arts institution, Mount Allison University, in Sackville, N.B. However, the unincorporated, loosely connected community from Lower Sackville through Middle Sackville to Upper Sackville, some 20 kilometres north of the centre of Halifax, was actually meant. The unofficial name, Sackville, has been gaining acceptance for this whole complex of some 25,000 people. While Nova Scotians may have no difficulty distinguishing their place from the little town of 5,500 people just over the border in the neighbouring province, the same name for two towns in the Maritimes will surely be confusing for other Canadians.

One of the most duplicated names in Canada is Windsor. As well as the city in southwestern Ontario, there are incorporated towns of Windsor in Québec and Nova Scotia, and an unincorporated community in New Brunswick. In Newfoundland, the town of Grand Falls–Windsor has a population of 14,700.

In 1973, the Ontario town of Durham in Grey County fought unsuccessfully to prevent the use of the same name for the new Regional Municipality of Durham, which embraces the county of Durham and parts of the counties of Ontario and Northumberland east of Toronto. Ontario has several other towns and villages whose names are duplicated by counties elsewhere in the province: Dundas, Simcoe, Perth, Prescott, Hastings, and Wellington.

The discrepancy of having the original community of Pickering, Ont., within the municipal boundaries of the town of Ajax and not those of the town of Pickering was partly resolved in 1987 by having it renamed Pickering Village. In 1975, the old community of Caledon, 50 kilometres northwest of Toronto, was renamed Caledon Village to distinguish it from the town of Caledon.

There are some confusing duplications in Western Canada too, starting with the largest city on the west coast, Vancouver. In 1886, it was incorporated as a city, duplicating not only the name of the biggest island on the coast, but also the name of a city in the state of Washington.

Other confusing community names in the west include: Hudson Bay in Saskatchewan; Dawson Creek in British Columbia and Dawson City in the Yukon; Sidney in British Columbia, often spelled the same as Sydney,

N.S.; Cochrane in both Alberta and Ontario; and Fort Saskatchewan, an Alberta town 25 kilometres northeast of Edmonton.

Although duplicated names will always be with us (Québec – city and province; Kenora, Ont., and Canora, Sask.; Chatham and Dorchester in both Ontario and New Brunswick), there are really not too many confusing look-alikes in Canada. Fortunately, for populated places with similar names, the correct postal code can usually prevent mail from going to the wrong look-alike.

Standard Spelling Eliminates Confusion ... Usually

Almost every place name in Canada has a single correct form, but some, like St. Catharines and Edmundston, are commonly misspelled. Some names of features crossing provincial boundaries (Timiskaming/Témiscamingue) or international boundaries (Kootenay/Kootenai) have variant spellings in the different jurisdictions. And a few names are spelled differently by postal or rail authorities.

One of the most frequently misspelled place names is St. Catharines, Ont. When I saw this spelling on a railway station sign in the 1950s, I assumed it was a mistake. Surely, I thought, there is only one proper way to spell St. Catherines. This assumption was reinforced by a sign pointing to St. Catherine's Cathedral.

The confusion about St. Catharines is compounded by a dispute over the origin of the name. The usually accepted story is that it refers to the first wife of Robert Hamilton, Catharine Askins Robertson, who died in 1796. A reference the same year to a church at 'St. Catherines' led Catharine Welland Merritt to claim in 1926 that the honour belonged to Catherine Butler, wife of the leader of Butler's Rangers. In doing so she chose to ignore an 1875 biography of her own grandfather, William Hamilton Merritt, which stated that the community was formally named in 1809 for the wife of Robert Hamilton, a prominent public official in the Niagara Peninsula in the late 1700s and early 1800s.

Another suggestion made in the 1920s that the name was given in honour of Catharine Prendergast, who married William Hamilton Merritt in 1815, has no substantiation. All that may be said is that both 'Catherine' and 'Catherine' were names commonly given to girls in the late 1700s, a time when Empress Catherine the Great of Russia (1729–1796) held court on the world stage.

Another name frequently misspelled is Edmundston, N.B., named in 1851 for Sir Edmund Walker Head, lieutenant-governor of New Brunswick from 1848 to 1854. Numerous references – maps, CN express slips, road signs, and advertising items – refer to 'Edmunston.'

During a test for a professional position with the Department of Energy, Mines and Resources in 1976, nine applicants were asked to locate Edmundston on a map; seven of them wrote 'Edmunston.' The omission of the 'd' may be partially explained by its phonetic weakness in a series of four consonants.

The proper spelling of Athabasca is difficult to pin down. Strong cartographic, documentary, and local practice in the 1800s favoured Athabasca, the form adopted in 1882 for one of the four new districts of the Northwest Territories. In 1902, the Geographic Board of Canada changed the name of the district, lake, and river to Athabaska, in keeping with international rules of nomenclature requiring a hard 'c' to be spelled with a 'k.'

After Alberta was established as a province in 1905, usage there favoured Athabasca, especially for the town 120 kilometres north of Edmonton. But when the Geographic Board of Canada was asked by postal officials in 1912 to review its decision, it reaffirmed the 'k' spelling it had made ten years earlier.

The sinking of H.M.C.S. *Athabaskan* during the Second World War, with great loss of life, reinforced the spelling 'Athabaska.' In 1947, however, the newly formed Geographic Board of Alberta adopted the form Athabasca and urged the federal board to concur. On receiving the agreement of Saskatchewan, in which more than half of Lake Athabasca is located, the Geographic Board of Canada agreed to Athabasca in 1948 for the river, lake, town, falls, mountain, and glacier, and several other minor features in Alberta.

Confusion about the spelling continues today, although prominence in the news of the Athabasca tar sands has encouraged acceptance of the 'c' form, but it is somewhat offset by Athabaska Airways, established in the 1990s to connect northern Saskatchewan places with major Prairie centres.

On the north side of the Fraser River, 50 kilometres east of Vancouver, is the community of Whonnock, a name derived from the Halkomelem word for 'humpback salmon.' In 1939, the Geographic Board respelled it Whonock, after the name of a local First Nations reserve. Railway and postal officials also opted for the one 'n' form, even though predominant local use had been Whonnock for a hundred years. After a concerted campaign in the 1960s, the extra 'n' was restored in 1969.

Also in the Fraser Valley, 40 kilometres farther east, the city of Chilliwack and the surrounding township of Chilliwhack were spelled differently for many years. But after a poll in 1979, the two were united on 1 January 1980 as the District Municipality of Chilliwack, which became a city in 1999. The name is from a Salish tribe; its meaning is uncertain.

In the late 1860s, Governor Frederick Seymour of British Columbia decided that one means of distinguishing American and British territories was to create different spellings for names of some prominent rivers. He decreed that the names Okanagan, Kootenay, and Yahk should be used in British Columbia to contrast with their spelling south of the border: Okanogan, Kootenai, and Yaak. These distinctions remain.

The spelling of 'Temiskaming' has caused much confusion, especially since this is not one of the three official forms of that name. Temiskaming was formerly part of the name of the Ontario Northland Railway (Temiskaming and Northern Ontario Railway). This spelling is also commonly used in the names of businesses in New Liskeard, Haileybury, and Cobalt.

The territorial district and federal electoral district in Ontario are spelled Timiskaming. The county and the federal electoral district in Québec are Témiscamingue. The lake's name is officially Timiskaming in Ontario and Témiscamingue in Québec. Adding to the confusion, the Québec town on the Ottawa River, 60 kilometres northeast of North Bay, is Témiscaming. So much for ensuring standard spelling for names with the same origin!

In 1982, the New Brunswick government legislated a change in the spelling of the town of Shippegan to Shippagan, but made no correction to several other names with Shippegan: the territorial parish, bay, park, beach, channel, and harbour. Canada Post spelled the name Shippegan from 1850 to 1852, 1862 to 1900, and 1955 to 1982, when it adopted Shippagan. The post office was Shippigan from 1852 to 1862 and 1900 to 1955, a common form of the name for associated features in the 1800s, and the form still used by Canadian National. Confusing, isn't it?

Who's to Blame for Mistake Names?

Mistakes happen – and are sometimes corrected. It would have been the Meach Lake Accord if Québec's Commission de toponymie had not repaired a century-old error in 1982. The lake got its name from pioneer settler Asa Meech, but it was misspelled Meach on an 1890 map and it took years (and appeals from Meech's descendants) to get it changed.

There are countless such tales of Canadian place names altered through spelling mistakes, bad handwriting, faulty translation, or confusion between sound-alike words. Often these alterations become commonly used and then prove very difficult to correct. Indeed, until about 1970, names authorities were reluctant to correct past errors at all, citing long usage in public documents, land descriptions, and maps. In recent years, however, serious requests to correct the spellings of names have been greeted with more sympathy, as in the case of Meech Lake (see pp. 225–8).

In 1765, Monckton Township was established at The Bend on the Petitcodiac River in New Brunswick, and named for British general Robert Monckton (1726–1782), who was lieutenant-governor of Nova Scotia from 1755 to 1759, and James Wolfe's second-in-command at the capture of Québec in 1759. The following year, when New Brunswick's townships were redesignated parishes, the 'k' was accidentally dropped. As a village grew at the 'Hub of the Maritimes,' it became known as Moncton, which was incorporated as a town in 1855, and a city in 1890. In March 1930, the city council passed a motion to rename the city Monckton, but fierce opposition ensued, and the motion was quickly rescinded in April.

During the early 1600s, a small river in Nova Scotia's

Annapolis Valley was named for Louis Hébert, the apothecary who accompanied Pierre du Gua de Monts to Port Royal, and Samuel de Champlain to Québec. When the Loyalists arrived in the late 1700s, they anglicized the name to Bear River. And that, to this day, is the name of the river, and a village of 900, nestled in a picturesque and surprisingly rugged setting, 12 kilometres inland from Digby.

Bury Head, near Alberton, P.E.I., was Barry Head on a 1770 plan and Berry Head on charts and maps from 1845 to 1966, when it was respelled to recall the site of an early graveyard. Some long-entrenched mistakes, on the other hand, are unlikely ever to be changed. Carleton Place, Ont., was named in 1830 for Carlton Place, a street in Glasgow, with its erroneous spelling probably influenced by nearby Carleton County. Aquaforte, Nfld., was initially Aguafuerte, but mapmaker Robert Robinson changed the Portuguese original on a 1669 map and the change stuck. And Geary, N.B. (pronounced 'GA-ree'), was settled about 1810 by Loyalists who called their settlement New Niagara for the Ontario peninsula where they had lived for a few years. Adopting the last two syllables of the contemporary pronunciation, 'NY-a-GA-ree,' they shortened the name to New Gary, and then to Geary.

Some incorrect names have been accepted without complaint. Around the turn of the century Joseph Chowaniec moved from Ledoux, Minn., and settled north of Weyburn, Sask. When his neighbours accepted his proposal in 1905 to name their new post office Ledoux, Chowaniec submitted it to the Post Office Department in Ottawa. A clerk misread his writing, however, and put Cedoux on the date stamp. It is Cedoux, Sask., to this day.

Cedoux is just one of many place names misspelled by government clerks and surveyors. On the Canadian National line, 30 kilometres northeast of Calgary, is the community of Kathyrn, named after Kathryn McKay, whose father donated the town site in 1911. The letters *r* and *y* were transposed on a sign in 1913, and the Post Office Department continued the erroneous – but by then adopted – form in 1919. As if to complete a circle, two Alberta place-names books published in the 1970s and the 1980 *Canada Gazetteer Atlas* all missed the long-established spelling of the community, referring to Kathryn rather than Kathyrn.

Philip and Helen Akrigg, authors of *British Columbia Place Names* (1986), have their summer home at Celista, B.C., on Shuswap Lake, northeast of Kamloops. In their book, they note that in 1908 postal authorities suggested Celesta, the name of a creek 35 kilometres to the northeast near the head of Seymour Arm, as the name for a post office. But an error in handwriting resulted in Celista. It became the name of the community, and was later adopted for the names of the creek and nearby Celista Mountain.

One of the main tributaries of the Ottawa River is the Madawaska River, which rises in Algonquin Park and joins the Ottawa River at Arnprior. Its name derives from the Matouoüescarini, a tribe encountered by Champlain in 1613, and it was called the Mataouachita on a 1703 map by Guillaume Delisle, and Mataouaschie in Sir David W. Smith's *Topographical Description of Upper Canada* in 1799. If the river's name had evolved solely from its rendering in early cartographic and documentary sources, its spelling might have become more like Matawashita. But its present spelling may have been influenced by New Brunswick's unrelated Madawaska River, which means 'land of the porcupine' in Maliseet. As early as 1725, engineer Gaspard Chaussegros de Léry spelled the Ontario river Matouasca, almost the exact spelling, Madouasca, used for the New Brunswick river on Delisle's 1703 map. In Alexander Sherriff's 1837 report of the lands between the Ottawa River and Georgian Bay, map maker William Henderson rendered the name Madaouaska. The present spelling was settled in surveyors' reports in the 1840s and on Thomas Devine's *Government Map of Canada* in 1861.

Two places in Ontario's Madawaska Valley have curious 'mistake' names. Camel Chute, at the head of Centennial Lake, 70 kilometres upriver from Arnprior, is called after a local sawmill owner named Campbell (a name sometimes pronounced 'KAM'l' in Eastern Ontario). When surveyor Duncan McDonell visited in 1846 he wrote Campbells Chute in his notes. When another surveyor, John Haslett, visited the next year, he first wrote Camels Chute in his diary but corrected it the following day to Campbells Chute. In 1890, the post office at the site was named Dubreuil, but difficulty with its spelling and pronunciation led local settlers in 1903 to ask for a name change – to Camel Chute. The MP for Renfrew South, Aaron A. Wright, mildly objected to the misspelling, but thought his protest of no consequence, 'since Camel Chute is shorter, it might be better to spell it that way, particularly since they have petitioned for it.'

Fifty-five kilometres farther up the Madawaska River from Camel Chute, on Kamaniskeg Lake, is the village of Barry's Bay. For a century and a half, local residents, historians, and names experts have puzzled over who the Barry was who left his name on the bay. In my *Geographical Names of Renfrew County* (1967), I mentioned the early name of 'Barry's Camp on the Bay.' And in the second edition (1989), a number of other claims were offered: Barry was a foreman for McLachlin Brothers, an Arnprior-based lumber company; he was a hunter or trapper; there wasn't a Barry at all but a field of blueberries!

But then, in 1992, I found in the Archives of Ontario the

misfiled microfilm of surveyor John Haslett's diary of his 1847 survey east from Kamaniskeg Lake and back again. Going east, he crossed 'Barry's Bay'; returning west, he crossed the Madawaska at lumberman William Byers's hay farm. As there were no settlers in the area before the late 1850s, I have concluded that Haslett wrote Barry's Bay when he should have written Byers's Bay, and the error was continued on maps and in survey records thereafter. It may be sheer coincidence, but even today Barry's Bay is often pronounced as though it were Byers Bay.

Acronyms ... Such as Kenora, Koocanusa, and Snafu Creek

Acronyms are words formed from one or more first letters of other words. Many do not reveal that they are not original words in their own right, such as Fiat, Sabena, and snafu.

Place-name specialists have been more lenient in their interpretation of 'acronym,' allowing names compounded from any parts of other words to qualify; for example, Mantario, a village in Saskatchewan, from Manitoba and Ontario.

Acronyms as place names are rare in Eastern Canada, but they have attained full flower in the Prairie provinces. There are at least fifty-six in Alberta, fifty-one in Saskatchewan, and thirty in Manitoba. Few are known beyond their own neighbourhood, and many have disappeared altogether, though the names persist in the old records.

Perhaps the best known of Canada's acronyms is Kenora, a city of some 10,000 inhabitants in northwestern Ontario.

In the late 1800s, the area on the north side of Lake of

the Woods was known as Rat Portage. In 1880, the postal designation Keewatin was chosen to replace that of Rat Portage, and in 1886 another post office on the north shore of the lake was named Norman for a son of a manager of the M & O Lumber Co. In 1904, a proposal was made to unite Keewatin, Norman, and the eastern part of Rat Portage into a single municipality known as Kenora. The name Kenora was adopted, but Keewatin remained a separate incorporated town until 1 January 2001, when it was united with the towns of Kenora and Jaffray Melick to form the city of Kenora.

As well as Mantario, Sask., Alsask, Sask., is a village near the Alberta border and northwest of Mantario, and Altario, Alta., is northwest of Alsask. Mankota, Sask., is a village, 50 kilometres north of the 49th parallel, whose first settlers came from Manitoba and North Dakota in the 1920s.

Another well-known acronym is Arvida, once a town in its own right, but since 1975 a neighbourhood of Jonquière in the Saguenay–Lac St-Jean region of Québec. Arvida was established in 1926 by the Aluminum Co. of Canada, and named for its president, Arthur Vining Davis.

Batawa, 7 kilometres north of Trenton, Ont., is derived from the Bata Shoe Company, plus an ending common to many Ontario names, such as Ottawa. Nolalu, 45 kilometres southwest of Thunder Bay, Ont., is named for the Northern Land and Lumber Company. Noranda, part of the Québec city of Rouyn–Noranda, is a contraction of North Canada.

When the Libby Dam was built in the state of Montana in 1967, a huge reservoir resulted, backing up the Kootenay (Montana's Kootenai) River into British Columbia. People and municipal authorities on the Canadian side favoured Lac Morigeau (after a pioneer). The preference of Montana residents, Koocanusa Lake, was determined to be the overwhelming favourite. In 1968, the United States and Canadian authorities approved Koocanusa, an acronym formed from Kootenay, Canada, and U.S.A.

Estevan, a city of some 9,000 people in southern Saskatchewan, derives its name from parts of the names of George Stephen and William Van Horne, the first two presidents of the CPR. North of Yorkton, Sask., is the town of Canora, named in 1905 from the first and second letters of Canadian Northern Railway.

Among Alberta places deriving their names from other words are: Cadomin (Canadian Dominion Mine), Marwayne (W.G. Marfleet and Wainfleet, England), and Westlock (two settlers, Westgate and Lockhart).

Two of Manitoba's better-known acronyms are Transcona (Transcontinental Railway and Strathcona) and Sherridon (Sherritt and Gordon).

The names Snafu Creek and Tarfu Creek in the Yukon are derived from acronyms of Second World War vintage. The first means, 'Situation normal, all fouled up'; the second, either 'That's always really fouled up' or 'Things are really fouled up.'

Use of acronyms as place names has been generally out of favour since the late 1960s. Provincial and territorial names authorities have usually rejected the cute and casual proposals submitted to them, preferring instead names having local or historical significance.

Avoid the Apostrophe ... But Not Always

The name Hudson's Bay has a long history, dating from the ill-fated voyage into this great body of water by Henry Hudson in 1610, and the establishment of the Governor and Company of Adventurers of England in 1670.

In 1900, the Geographic Board of Canada published the following decisions: 'Hudson bay and strait; inland sea and passage communicating with the Atlantic (not Hudson's).' Canadian maps and charts then began using the name Hudson Bay. Subsequently, the Hudson Bay Railway was built to Churchill, Man., the Hudson Bay Mining and Smelting Company was opened in Flin Flon, Man., and the town of Hudson Bay (pop. 1,900) was established in Saskatchewan, east of Prince Albert.

The board's decision was in keeping with one of its basic rules adopted in 1898, which stated that the possessive form of names should be avoided whenever it could be done without destroying the euphony of a name or changing its descriptive application. The rule added that, if the possessive were retained, the apostrophe should be dropped.

This rule had universal application in the English-speaking world, having been adopted in the 1890s by names authorities in Britain and the United States (while respecting well-established names with the apostrophe), and later in the twentieth century by Australia and New Zealand. But in the 1970s, the rule was amended in Canada to permit the apostrophe where it was well established and in current use.

There are several names in Canada with the apostrophe. Examples are St. John's in Newfoundland, St. Peter's in Nova Scotia, Campbell's Bay in Québec, and Lion's Head in Ontario. These forms are in keeping with the principle of geographical naming that names established in the statutes by other authorities must be accepted without change.

In the first half of the twentieth century, the Geographic Board severely restricted the use of the possessive 's,' even when local communities had long used it – such as Reed's Point, Moore's Hill, and so on.

Occasionally, in its zeal to trim the 's' from names, the board erroneously modified some personal names ending in 's.' In Lanark County, Ont., the board imposed Wood Lake, Nichol Lake, and Bower Lake even though these were derived from the family names Woods, Nichols, and Bowers. Corrected forms were entered into the official records in the 1960s.

The practice of omitting the apostrophe from supposed possessive names varies from community to community. The town of Smiths Falls, Ont., was established in the statutes in 1882 with the apostrophe, but that spelling was ignored locally. The Ontario Municipal Board noted the discrepancy in 1968, and required the Ontario legislature to pass an act recognizing the form Smiths Falls before allowing the issuing of debentures. The town of St. Marys, Ont., similarly did not use the apostrophe, although recorded in the statutes, so it was also changed to its current form in 1968. Subsequently, the town authorities persuaded the St. Mary's Cement Company, whose trucks are seen throughout the province, to omit the apostrophe from its name.

The Geographic Board did not accept the statutory name Lion's Head, Ont., in 1913, but substituted Lionhead. The form Lion's Head, widely used for local services and the post office, was restored in 1940. Nearby is the local community of Colpoy's Bay. The adjacent bay (on the Georgian Bay side of the Bruce Peninsula) was named about 1820 for Sir Edward Colpoys, but the local residents, unaware that the terminal 's' is not possessive, use an apostrophe. The name Barry's Bay, Ont., is used by the residents, but the post office is Barrys Bay. The name Burk's Falls, Ont., is also used widely, although it is ignored by the postal authorities and the weekly newspaper, the *Burks Falls–Powassan Almaguin News*.

The apostrophe is rarely used in official names in

western and northern Canada. Some writers refer to Gibson's Landing, a charming community on Howe Sound northwest of Vancouver, where the popular TV show *The Beachcombers* was based, but the village has long been known as simply Gibsons. The municipality of Hudson's Hope in northeastern British Columbia was established in 1965, but the post office, opened in 1913, remains Hudsons Hope.

The Lions Gate Bridge and the village of Lions Bay both derive their names from 'The Lions,' distinctive peaks that look down on Vancouver. Some writers incorrectly add an apostrophe either before or after the 's' in such names.

Names like Kingsport, Kingscourt, and Kingston do not require the apostrophe, although they are composed of two or more fused words, with the prefix suggesting a possessive characteristic. The same approach should be followed with names such as Kings Point, Kings Lake, and Kings Head since, as proper names of features in the public domain, they no longer have any possessive connotation. It follows that names like McLarens Landing, Fiddlers Elbow, and Haydens Corners also do not require apostrophes.

In dealing with names ending in 's' with the possessive apostrophe, it is wise to avoid autocratic rulings. If residents insist on using St. Kyran's, Heart's Content, Colpoy's Bay, and Ayer's Cliff, names authorities should be flexible enough to grant them their wish.

A New Name for Merged Municipalities Can Cause Quite a Fuss

When well-established municipalities are being forced to amalgamate and select a new name, without adequate consultative processes, much discord and tension are likely to follow.

Shakespeare wrote in *Romeo and Juliet* that 'a rose by any other name would smell as sweet,' but the twin Ontario cities of Fort William and Port Arthur weren't buying it back in 1968 when the province decreed they could no longer exist as separate entities. These historic places at the head of Lake Superior had been about equal in area and population for most of the seventy-five years before that, and developed a strong rivalry over which name the new city would bear.

To end the wrangling, the minister of municipal affairs, Darcy McKeough, asked the Lakehead Planning Board to ignore all existing names and devise some alternatives. He subsequently allowed only the board's three suggestions to be put to the voters. Thunder Bay, from the historic bay bordering the sparring municipalities, received 15,821 votes; Lakehead, 15,302; and The Lakehead, 8,477. Clearly, the majority of the voters preferred some form of Lakehead, but their vote was split. It was not long, however, before Thunder Bay was fully accepted, while Lakehead survives in the identities of various businesses and institutions, including Lakehead University. Even Fort William and Port Arthur linger on, albeit in informal usage by older residents and in the names of provincial electoral divisions.

Finding a compromise name with a strong historical association proved effective in New Brunswick as well. On 1 January 1995, the towns of Newcastle and Chatham and adjacent villages on both banks of the Miramichi River became a single city under the name of Miramichi. There was a similar development in southern Alberta when the well-known local name Crowsnest Pass was given to the new municipality that emerged from the fusion of the towns of Blairmore and

Coleman, and the villages of Bellevue, Frank, and Hill-crest Mines.

History played no part, however, in resolving the con-flict in Ontario when the town of Bowmanville, the vil-lage of Newcastle, and the townships of Clarke and Darlington were united in 1974 as the town of Newcas-tle. Although most voters did not know it, the name Newcastle was deeply rooted in the history of that area of the north shore of Lake Ontario, where the district of Newcastle had existed from 1799 to 1850. Many assumed that the little village had been given precedence over the town of Bowmanville. Confusion and public unrest con-tinued for fifteen years until, in 1991, a name change was endorsed by 59 per cent of the voters. Eventually, after extensive consultation, a committee proposed the name Clarington for the new 'municipality,' which the resi-dents wanted it to be called, deeming 'town' to misrep-resent the nature of the largely rural community. Clarington, a blend of Clarke and Darlington, has more than 58,000 people, but the name is largely unknown outside the area of Durham Region.

Merger mania reached a high level in Ontario in the 1960s and '70s, and attained super megalomania in the province in the 1990s. Amalgamations were promoted in Québec, Newfoundland, and New Brunswick in the 1970s and '80s. The process got started in Prince Edward Island and British Columbia in the early 1990s. There has been little need for large-scale municipal streamlining on the Prairies, although Winnipeg annexed several subur-ban municipalities in the 1960s, and many cities, espe-cially Calgary and Edmonton, have extended their boundaries to encompass outlying suburbs.

Economic reasons, such as cost efficiency and improved delivery of programs, usually lie behind municipal amal-gamations. Sometimes they evolve through local consent, but more often they come about through provincial enforcement. In joining together two or more municipalities, important questions of identification arise: what to call the new municipality; how to choose its name; what to do with former names? Mergers can be highly emotional experiences, especially for long-time residents who cherish their heritage. Arriving at an acceptable name calls for tact, sensitivity and in-depth knowledge of local history.

To be sure, some names are naturals. When one municipality is considerably larger than the others entering the union, for instance, its name invariably takes precedence, with the smaller municipalities being remembered as neighbourhood and ward names. This happened in St. Catharines, Ont., in 1961, when the city absorbed the towns of Merritton and Port Dalhousie; and in Jonquière, Qué., with the annexation of the towns of Arvida and Kénogami in 1975. When the more populous district of Abbotsford, B.C., was united in 1995 with the district of Matsqui, Abbotsford became the new city's name. During the same year the cities of Halifax and Dartmouth, the town of Bedford and the county of Halifax were joined together as the Regional Municipality of Halifax. In 1996 the government of Ontario decided, without full consultation of the residents of Metropolitan Toronto, that Toronto would become an enlarged city on 1 January 1998. The cities of Etobicoke, North York, Scarborough, and York, and the borough of East York were absorbed into the new city of 2,385,421 people. The names of the former municipalities continue to have relevance to their respective residents.

In the merging of two municipalities with nearly equal areas and populations, the new name is often a hyphenated combination, as in Grand Falls–Windsor, Nfld., joined in 1991, and Rouyn–Noranda, Qué., amalgamated in 1986.

Even with hyphenated names, one of the names may eventually supersede the other. The New Brunswick town of St. Stephen–Milltown, amalgamated in 1973, became simply St. Stephen two years later. Similarly, Lévis–Lauzon, Qué., united in 1989, was replaced by Lévis alone in 1991 when the town of Saint-David-de-l'Auberivière was brought into the fold. Other examples are Victoriaville, Qué., after being Victoriaville–Arthabaska for less than a year, and Sarnia, Ont., which was officially Sarnia–Clearwater during 1990 and 1991.

Sometimes the choice of a new name has been found in an obscure connection. In 1972, a row erupted in the Ontario towns of Galt, Hespeler, and Preston, as well as parts of the adjacent townships, when Galt was designated by the province as the name for the planned amalgamation of these municipalities. So strong was the ensuring protest that all existing names were promptly disqualified, and the province called for a completely new name. The Preston and Hespeler councils agreed on Cambridge, after Cambridge Mills, an early name of Preston, while Galt's council settled on Blair, the name of a community on the west side of the Grand River. By a narrow margin – 11,728 votes to 9,899 – Cambridge won. Since then, Preston and Hespeler have remained strong neighbourhood names, but Galt has faded somewhat in use since it became the main urban centre of Cambridge.

In 1993, the municipal communities of Southport, Bunbury, Crossroads, and Keppoch–Kinlock, in Prince Edward Island, were united as the town of Waterview, but that name did not sit well with some vocal opponents, who demanded a second ballot. The next vote supported Stratford, the second-place finisher in the earlier vote. That name had been derived from the main road in Southport, which had been first used in 1858 for the name of a local school district.

New names may also be cobbled together out of former names. Walden, west of Sudbury, Ont., was a 1973 concoction of parts of the names Waters and Denison townships, and the town of Lively. The fact that it replicated the name of the pond in Massachusetts made famous by Henry Thoreau was no doubt not far from the thoughts of local councillors. On 1 January 2001, the town of Walden was absorbed by the new city of Greater Sudbury. Ramara Township in Ontario's Simcoe County grew out of the union of Rama and Mara townships in 1994. Many of the recent amalgamations in Ontario have resulted in some of the dullest names ever foisted upon the province's municipal toponymy, examples being South Bruce Peninsula, Northeastern Manitoulin and the Islands, and North Frontenac.

As the benefits of 'bigger is better' have increasingly appealed to cash-strapped governments, more and more mergers will take place down the road. In the area of Canada's national capital the Regional Municipality of Ottawa–Carleton, the cities of Ottawa, Nepean, Gloucester, Kanata, Cumberland, and Vanier, the village of Rockcliffe Park, and the municipal townships of Osgoode, Rideau, Goulbourn, and West Carleton became the city of Ottawa on 1 January 2001. The twenty-nine municipalities of the Île de Montréal may be merged under the single name of Montréal, but many of the present officially bilingual cities may insist on protecting in law their linguistic status within the new super city.

When Names Become Politically Incorrect

Pride and fond memories of homelands and heroic leaders across the sea have influenced many place names in Canada. London, Ont., New Glasgow, N.S., New Denmark, N.B., and Labrador's Churchill River are just a few of the hundreds of examples. However, several name changes took place after Canadians fought wars against countries where some of our names originated, or when foreign heroes were subsequently judged monstrous.

Of the three international powers Canada has been at war with in the twentieth century – Germany, Italy, and Japan – only Germany has been the source of a large number of Canadian place names. By the beginning of the First World War, there were some thirty names of probable German origin in southwestern Ontario between Stratford and Guelph – the largest concentration of German settlement in Canada. Only three of these were changed during wartime. In 1942, a small community between Guelph and Waterloo, originally called New Germany, was renamed Maryhill after the Blessed Virgin Mary and the hill on which St. Boniface Church was located. Two years later, German Mills – 6 kilometres south of downtown Kitchener – was changed to Parkway.

The most significant change in the area occurred during the First World War, when the city of Berlin – now Kitchener – was forced to choose a name more in tune with the ideals of freedom, democracy, and service to the British Empire.

Berlin was named in 1833, and by 1911 it was the ninth-largest city in Ontario, with 70 per cent of its population having German roots. But by 1916, the name Berlin had become untenable, observed co-authors John English and Kenneth McLaughlin in *Kitchener: An Illustrated History.* Berlin's city council and business leaders feared the patriotism of its residents would be judged suspect, and worried about the loss of business. In February 1916, the council appealed to the Ontario legislature for permission to change the name, but it was told to hold a plebiscite first. On 19 May, 3,057 Berliners cast their ballot, with approval for a name change winning by only eighty-one votes. But the change to Kitchener was not made without considerable resentment and acrimony, especially in the German parts of the city.

The selection of a new name was not easy. From thousands of submissions, the choice was reduced to six, all of which were rejected by council because of widespread criticism. Council also defeated a subsequent proposal to amalgamate with Waterloo, and instead drew up another list of six names: Adanac, Benton, Brock, Corona, Keowana, and Kitchener. The last was in honour of Lord Herbert Kitchener, a British military hero who drowned on 5 June 1916. Only 892 residents voted, with Kitchener receiving the most votes, 346, followed closely by Brock with 335. The remaining four received only 48, with 163 voters spoiling their ballots by writing in either Berlin or Waterloo. Considering the turnout too low to justify a change, opponents gathered twice as many signatures in opposition. Regardless, council changed Berlin to Kitchener on 1 September 1916.

Not all efforts to change names have succeeded. In one case, a place name with Nazi overtones – Swastika, Ont. – was successfully protected from assault by toponymic purifiers. A mine was named in 1906 by Bill and Jim Dusty, who discovered the first gold in Northern Ontario while exploring for silver on behalf of investors in Tavistock, Ont. A cross on a charm worn by a woman from Tavistock inspired Bill Dusty to name the mine Swastika, meaning 'good luck' in Sanskrit. Both the community and railway station adopted the name.

But during the Second World War, the swastika – the official emblem of the Nazis – came to symbolize misery and the annihilation of millions. Consequently, Ontario's Minister of Highways announced in 1940 that Swastika would be changed to Winston in honour of the British prime minister, Winston Churchill. In defiance, residents of Swastika tore down the new sign, replacing it with another that read: 'To Hell with Hitler. We had the swastika first.' As John Wroe reported in *Saturday Night* magazine in 1986, 'the government never got its wish and, as the war receded in memory, the controversy diminished.' Since then, occasional efforts by outsiders to get rid of the name have been fiercely resisted by the residents of Swastika, now part of the town of Kirkland Lake.

In 1936, a Canadian Pacific station across the Fraser River from Hope, B.C., was named Petain in honour of Marshal Henri Philippe Pétain, a distinguished French military leader during the First World War. Four years later, when the free world learned that Pétain had collaborated with Nazi Germany, Canadian Pacific changed that station name to Odlum for Maj. Gen. Victor Odlum, then a widely respected Canadian diplomat. But Mount Pétain in the Canadian Rockies and two lakes and a township in Québec remain as honours to the French military genius.

Military and political leaders who fought with the Allied powers during the Second World War have also had their names affixed to geographical features in Canada. Among those honoured was Marshal Joseph Stalin, whose name was given to a mountain in north-central British Columbia and a township 50 kilometres southwest of Sudbury. But in 1985, names scholar Jaroslav Rudnyckyj and members of the Ukrainian Canadian Committee urged federal and provincial politicians and names boards to expunge the name Stalin – a person they deemed to be one of history's most notorious criminals – from Canadian maps. In the Ontario legislature, the cause was taken up by MPP Yuri Shymko, whose private member's bill to substitute Hansen Township for

Stalin Township was unanimously passed on 6 November 1986. The new name honours Rick Hansen, who gained international attention during his Man in Motion wheelchair tour to promote research into spinal cord injuries and other physical disabilities.

In April 1987, the British Columbia government followed Ontario's lead, changing Mount Stalin to Mount Peck in honour of Don Peck, a widely respected trapper, guide, and outfitter who lived at Trutch, B.C., on the Alaska Highway, from 1949 to 1963. But the change was not universally endorsed. The Vancouver Historical Society protested the rewriting of an important part of Canadian history. And one Fort Nelson resident wrote to several newspapers to say that Stalin had been honoured because he stood steadfast in defeating the German foe, and that to eliminate his name from our maps was no better than the Stalinesque purges impugned by the free world.

Near Canmore there is a mountain that had long been known as Chinamans Peak, which recalled a climbing feat performed in 1896 by miner Ha Ling, whose ancestry was Chinese. One hundred years after the climb, several residents of Canmore and Calgary expessed their displeasure with what they perceived to be a derogatory name. In 1998, the names authority in Alberta had it renamed Ha Ling Peak, a recommendation that had been made by the *Medicine Hat News* in its 'Canmore Cullings' column in 1896.

Some names now gracing our cities, towns, mountains, and rivers may well be judged politically incorrect in the years to come, and vigorous efforts may yet be made to replace them with names considered more appropriate to Canada's vision of peace, tolerance, and freedom. However, since I am not a revisionist at heart, I hope such demands for change will be rare.

Hot and Bothered by 'Disgusting' Names

Place names derived from words deemed coarse or unspeakable have frequently aroused sensitivity, offence, and even hostility. Over the years new names and masked substitutes have been used to banish names found vulgar or suggestive.

In 1924, almost 200 property owners in Victoria asked city council to change the name Foul Bay to Gonzales Bay, after Gonzales Point, named in 1790 for the first mate of a Spanish sloop. The change was opposed by Post No. 3 of the Native Daughters of British Columbia, among others, but the matter was raised again ten years later by Henri Parizeau, long-time director of the Canadian Hydrographic Service on the west coast. He claimed that 'it would be in the interests of the public in general, that this most objectionable name around the coast of Victoria should be changed to a more respectable and decent name.' It was a sentiment shared by the Foul Bay Community Association, which noted that the name was 'a constant source of offence to residents and of ridicule by non-residents and tourists,' as well as an obstacle to residential development. Finally, in November 1934, with endorsements from the councils of Victoria and Oak Bay and provincial authorities, the Geographic Board of Canada authorized the change. However, the boundary street between the two municipalities is still Foul Bay Road.

Two names of Native origin on Vancouver Island, Kokshittle Arm and Kowshet Cove, were also changed in 1984, to Kashutl Inlet and Cullite Cove. George G. Aiken, provincial member of the Geographic Board, assumed a high moral tone in support of the changes. 'We do not wish,' he wrote, 'to have the daughters of our present and future citizens feel embarrassment in naming the locations of their homes. We cannot, as a Board, in this democratic country, ignore these objections of the citizenry when these objections are founded on good taste and reason.'

In 1959, a family of Finnish extraction named Suni advised the Ontario Department of Municipal Affairs that the word *paska* ('shallow' in Cree), as in the name Paska Township, was a vulgar word for excrement in Finnish. Municipal Affairs advised the Department of Natural Resources of the offending names Paska Lake and Paska Creek, and Canadian National Railways of the embarrassingly named Paska siding. As there was a large Finnish population in Northern Ontario, especially in Thunder Bay, 250 kilometres to the southwest, the township, the siding, and the two natural features were renamed – after the Suni family.

Other examples of undesirable names changed include, in Newfoundland, Distress becoming St. Bride's, Famish Gut switching to Fairhaven, Scilly Cove opting for Winterton, and Turk's Gut changing to Marysvale. In Alberta, Meighen (pronounced almost like 'mean') was changed to Viking. And in New Brunswick, Pisarinco became Lorneville.

A century ago, Sipweske was a growing village near Brandon, Man. Its name, derived from a Cree word, had unwelcome overtones of intemperance to some. According to Penny Ham in her *Place Names of Manitoba* (1980), Wawanesa ('whippoorwill'), adapted from Wawonaissa in Longfellow's *The Song of Hiawatha*, was then introduced. The village of 482 is the headquarters of the Wawanesa Mutual Insurance Company.

Crotch Lake is a reservoir on Ontario's Mississippi River, 110 kilometres southwest of Ottawa. In 1848, land surveyor John Harper gave it the euphemism Cross Lake, and this was endorsed by the Geographic Board in 1941. I made a field survey of the area in 1965, and found the original name was universally preferred. It was subsequently reaffirmed by the Canadian Permanent Committee on Geographical Names.

In 1910, the Lethbridge Board of Trade declared, 'We are certainly very much disgusted with the present name and wish to make a change of some kind.'

The object of the board's disgust was the name of the river on whose banks the city of Lethbridge, Alta., was situated. It was the Belly River, and Lethbridge citizens were convinced that their thriving young city deserved to be set on a river whose name did not engender discomfort and embarrassment. (In those days 'belly' was a word avoided in polite company.) They had campaigned since 1886, but it took almost thirty years to secure a more distinguished appellation.

Alberta's Belly River rises some 20 kilometres south of the 49th parallel in Montana and flows north to join the Oldman River 18 kilometres upstream from Lethbridge. It was given its name during a survey in 1858 by Thomas Blakiston, a member of the Palliser Expedition, and the name first appeared on a map in 1865.

Blakiston derived the name from the Blackfoot *Mokowanis*, meaning 'big bellies,' referring to the local Atsina tribe and equivalent to the name Gros Ventre used by French explorers and voyageurs. Blakiston applied the name Belly to the whole river from its source to its confluence with the Bow River, where the two rivers become the South Saskatchewan River, 90 kilometres northeast of present-day Lethbridge. Thus when the city of Lethbridge was first laid out near a series of rich coal deposits in 1885, it found itself on the east bank of the Belly River.

Just one year later, Minister of the Interior Thomas White visited Lethbridge, and residents asked him to get the name changed. No action was taken, and in 1900, *Lethbridge News* publisher E.T. Saunders – who had earlier found the name historic and euphonious – urged a change, too. He wrote about the discomfort of strangers and the blushing embarrassment of ladies. To get around the problem, he suggested that the St. Mary River, a tributary of the Belly upriver from Lethbridge, be designated the main watercourse, with the Belly as a branch of it. That way, Lethbridge would sit beside the St. Mary.

Six years later (and still no action), Saunders again appealed to the authorities. He claimed the name Belly made Lethbridge appear to be a crude frontier settlement, and this time suggested extending the name South Saskatchewan River to some point upriver from Lethbridge.

In 1910, the Lethbridge Board of Trade wrote a letter to the Department of the Interior proposing the name Alberta River for the entire watercourse, including the present Oldman and South Saskatchewan rivers. The Geographic Board rejected the proposal, stating the new name had no authority based on usage, and noting in any case that the Bow was the principal tributary of the South Saskatchewan. The Board of Trade asked for other alternatives to Belly, but the Geographic Board responded by urging it to canvass local residents for their own alternatives.

The matter languished until 1913, when the Board of Trade again asked the Department of the Interior to replace Belly River with Alberta River, 'a name citizens could call attention to with pride.' But the Geographic Board told the Board of Trade to obtain the views of local authorities and then seek the endorsement of the provincial government. In 1915, the Board of Trade asked Premier Arthur Sifton to approve Alberta River, but Sifton said he preferred Lethbridge River. This new proposal was sent to the Geographic Board, which declined to accept it. Geologist and Geographic Board member

Donaldson Dowling pointed out that the Oldman River carried considerably more water than the Belly at their confluence, and suggested carrying the name Oldman River to its juncture with the Bow. This would make the Belly the Oldman's tributary and put Lethbridge on the Oldman. It was the compromise the Geographic Board approved in August 1915, and thus after many years of appeals, Lethbridge finally got rid of its Belly River.

Fortunately, the purging of place names has become quite rare in recent times. People are sometimes astonished to encounter Bastard Township in Ontario's Leeds County, but in 1796, when the township was surveyed and named, Bastard was a respectable surname in England, and John Pollexfen Bastard was a member of Parliament. The names of both Bastard Township and adjoining Kitley Township are derived from the Bastard family seat in Devon.

In 1966, I asked Reeve Gerald Cross of the Township of Bastard and South Burgess if thought had ever been given to changing the name. He replied that the question had been raised with the Department of Municipal Affairs in Toronto, but when told it would cost $1,200, the council decided such funds should be spent on more important matters. Local efforts by a realtor in 1994 to rename it Beverly Hills (after the Upper Beverly and Lower Beverly lakes), on the two-hundredth anniversary of the naming of the township, were rejected by the local residents. It is now part of the Municipal Township of Rideau Lakes.

Castle Mountain and Its Thirty-Five-Year Eisenhower Hiatus

On 17 August 1858, Dr. James Hector, a naturalist-geologist with the Palliser Expedition, was exploring the

Bow River valley upstream from present-day Banff. Beside Vermilion Pass (which now leads to Windermere, B.C.), he noted a majestic 11-kilometre-long mountain. It reminded him of the great crenellated castles of Britain, and he promptly named it Castle Mountain.

(Two days later, Thomas Blakiston, another naturalist with the same expedition, assigned the name Castle Mountain to a feature west of present Pincher Creek, which is about 250 kilometres southeast of Banff. However, it was renamed Windsor Mountain in 1915.)

On 9 January 1946, the day before the visit to Ottawa by Gen. Dwight Eisenhower, then Supreme Commander of the Allied Forces in Europe, Prime Minister Mackenzie King, acting on a suggestion from his friend Leonard Brockington, instructed the Geographic Board of Canada to rename Castle Mountain in honour of the illustrious general. Apparently the prime minister had learned that the general had been presented with a castle in Scotland, and Mr. King regarded the presentation of this Canadian 'castle' as a gift of much greater magnitude and significance.

In Alberta, the provincial government was astonished by the arbitrary name change, and proceeded to establish its own names board to make sure that all naming in Alberta henceforth would receive provincial approval. The Alpine Club of Canada called the renaming a sacrilege; and editorials noted that there were many unnamed mountains in the Rockies to which the general's name could have been given.

In the early 1970s, efforts were begun to restore the name Castle Mountain and to assign Eisenhower to another feature. Freeman Keyte, a public servant who was living in Calgary at the time, took the crusade to the Geographic Board of Alberta, members of Parliament, and the Canadian Permanent Committee on Geographical Names. The editor of the *Crag and Canyon* in Banff raised the issue with the provincial minister of culture, and subsequently newspapers in Calgary and Edmonton published many letters urging reinstatement of the name Castle Mountain. The letter writers claimed that the original name embodied the country's ruggedness and independence. The publisher of the *Edmonton Journal*, however, expressed his opposition to the reinstatement.

In February 1975, the Historical Society of Alberta passed a resolution calling for restoration of the name Castle Mountain, and suggested that Eisenhower Peak be given to its most prominent tower. This was endorsed by the Alberta Historic Sites Board, which had succeeded the Geographic Board of Alberta as the new provincial names authority.

Because the mountain is in Banff National Park, the matter was referred to Parks Canada, then part of the Department of Indian and Northern Affairs. The name change was approved, but no announcement was to be made until after the United States bicentennial celebrations in 1976. However, the news became known in April 1976 to the *Edmonton Journal*, whose publisher repeated his opposition.

In May 1976, the issue was raised in the House of Commons by former prime minister John Diefenbaker, then an opposition MP. Because of a perceived bond with Gen. Eisenhower through their common efforts to defeat the German forces during the First World War, Diefenbaker was strongly opposed to changing the name back to Castle Mountain. Alistair Gillespie, the minister responsible, assured the House that the name Mount Eisenhower would stay. And so it did ... for another three years.

However, immediately after the election of the Conservative government of Joe Clark in May 1979, the Alberta

Historic Sites Board took up the issue again. Freeman Keyte, then residing in the Ottawa area, pursued the matter tenaciously with Mr. Clark's office, with the minister responsible for Parks Canada, John Fraser, and with the provincial minister of culture, Mary J. LeMessurier.

Although federal endorsement of the change back to Castle Mountain was finally given in October, Alberta's concurrence was not received until the morning of Thursday, 13 December. On the evening of the same day, Mr. Clark's government was defeated in a non-confidence vote in the House.

This landmark federal–provincial decision restored Castle Mountain, and assigned Eisenhower Peak to its prominent eastern elevation. It also confirmed the name Castle Junction for the intersection of Highways 1A and 93.

The *Mount Eisenhower* 1:50,000 topographical map was scheduled for a new printing in the fall of 1979. I persuaded the chief of map reproduction to withhold the map until the name was changed. The *Castle Mountain* map was published in January 1980.

In the years since the reinstatement of Castle Mountain, the only reaction received by the committee's names secretariat has been that of enthusiasm and pleasure.

Receiving Names from Abroad and Exporting Names

Touring the Middle East without Leaving Canada

The Gulf crisis of 1990–1 focused the world's attention on the city of Baghdad, capital of Iraq since 1921. This ancient city of Mesopotamia lies in the fertile valley of the Tigris, only 40 kilometres east of the Euphrates River.

It is one of many historic Middle East cities, regions, mountains, and rivers reflected in Canada's place names. In the late 1880s, the name Bagdad (without the 'h') was given by the Central Railway of New Brunswick to a flagstop 35 kilometres northwest of Sussex. Although Canadian Pacific abandoned the line in the 1960s, Bagdad remains the name of the small community.

Another ancient city of the Tigris Valley recalled in Canada is Nineveh, once the capital of the Assyrian Empire. Its intricate streets, squares, and a magnificent palace were unearthed by British archaeologists between 1820 and 1851. The farming community of Nineveh, 25 kilometres northwest of Bridgewater, N.S., was settled in the 1840s. There is another locality by the same name on Cape Breton Island, 23 kilometres southwest of Baddeck.

Tales of the mysterious Middle East and stories from the Bible are likely the source of some 140 place names in Canada. About 75 of these occur in Southern Ontario, in an arc from Chatham in the west through Barrie and Newmarket to Belleville in the east. There are clusters of such names in a 40-kilometre radius around both Belleville and Lindsay. Québec, New Brunswick, and Nova Scotia have about a dozen each, with concentrations northeast of Granby, Qué., and around Sussex, N.B. In all of Western Canada, there are only about 15 Middle East names. Only one appears to have been given in the three territories: Ophir Creek in the Yukon.

The giving of Middle East place names in Canada appears to be a nineteenth-century phenomenon. Most were chosen between 1820 and 1885, when much of the naming took place in the Maritimes, Southern Québec, and Southern Ontario. Many towns and villages in Northern Ontario and the West were founded after 1885, when few turned to the Bible or Arabic tales for inspiration.

Many names from the Middle East were given to rural communities or crossroads where Protestant congregations established meeting-houses. When a school, post office, or railway flagstop was established, the leaders of each community encouraged the use of their church name.

The original Garden of Eden, thought to have been in the Middle East near the headwaters of the Euphrates, has also influenced Canadian place names. About 30 kilometres southwest of Antigonish, N.S., are Garden of Eden and Eden Lake. There is also a place called Eden 35 kilometres northeast of Port Hawkesbury, N.S. In Ontario, there is a Garden of Eden 10 kilometres north of Renfrew, an Eden south of Tillsonburg, and an Eden Grove northwest of Walkerton. Eden Mills is east of Guelph, and Edenvale is northwest of Barrie. There is one Eden in Manitoba, 15 kilometres north of Neepawa. In Saskatchewan, the village of Edenwold, 35 kilometres northeast of Regina, was a German settlement when it was named in 1890. Kaleden, B.C., was the winning name in a 1909 contest: a minister combined the Greek word for 'beautiful' with the name of the biblical garden of paradise.

Egypt and Lebanon are also among Canada's place names. Nova Scotia has a place called Egypt Road, and Québec has a place called Égypte. In Ontario, Egypt is just south of Lake Simcoe, Little Egypt is northeast of Goderich, and Lebanon is 35 kilometres northwest of Waterloo. The regions of Ophir and Sheba, in the southern part of the Arabian Peninsula, are both reflected in the names of townships in Northern Ontario. Ophir is also a community 50 kilometres east of Sault Ste. Marie. Sheba was once a Canadian Pacific flagstop northwest of Thunder Bay.

Cairo and Damascus also turn up in Canada. In 1852, a post office called Sutherland's was opened 40 kilometres northeast of Chatham, Ont. In 1896, it was changed to Cario, and respelled Cairo the following year. There is also a Cairo Township in Northern Ontario. Damascus, 30 kilometres west of Orangeville, Ont., dates back to 1872. Another Damascus, 28 kilometres northeast of Saint John, N.B., was named by a surveyor in jest because the tortuous road leading to it reminded him of Paul's conversion on the way to the original Damascus.

The biblical story of Moses leading the Israelites from the Land of Goshen (near the present mouth of the Nile) to the promised Land of Canaan prompted many early settlers to name their communities after one or the other. In the early 1800s, a traveller baptised two places south of Renfrew, Ont., Goshen and Canaan when he was denied food in the first, but was offered nourishment in the latter. Both names were adopted by their respective communities, but ironically only Goshen survived.

There are two places called Goshen in New Brunswick – one 25 kilometres northwest of Sussex, and the other 25 kilometres east of Sussex. To avoid confusion, the latter used the name Sheba for its postal address, but the community remained Goshen. The Canaan River in New Brunswick spawned several community names: New Canaan, Canaan Road, Canaan Station, and Canaan Rapids.

Nova Scotia has two places called Goshen. One is 65 kilometres northwest of Halifax. The other, 27 kilometres south of Antigonish, was so named because twice a year two members of the community were sent to Halifax for supplies, much like the sons of Jacob went to Goshen in Egypt to buy corn. There are also three places called Canaan in Nova Scotia, and another called New Canaan.

Before he died, Moses viewed the promised land from atop Mount Pisgah, a height of Mount Nebo east of Jericho and the River Jordan. Mount Pisgah, 13 kilometres northeast of Sussex, N.B., offers an excellent view of lush fields in the valleys below. Mont Nebo, 75 kilometres west of Prince Albert, Sask., was at one time a Canadian National flagstop.

Mount Hebron, Mount Hermon, and Mount Carmel are also prominently mentioned in biblical stories. Hebron in Labrador was established by Moravian missionaries in 1829. In New Brunswick, Mount Hebron is 10 kilometres north of Sussex. In Ontario, New Hermon is 20 kilometres east of Bancroft. Mount Carmel occurs as a place name five times in Ontario, and is the name of an Acadian community in western Prince Edward Island.

Mount Zion was the fabled city of David on the eastern ridge of Jerusalem. New Brunswick has New Zion and Zionville, and Ontario has six places called Zion, as well as Zion Hill and Zion Line.

Jerusalem itself was widely known as Salem. This name occurs six times in Ontario; there is also a Salem

Corners southwest of Lindsay. Salem, New Salem, and Salem Road can all be found in Nova Scotia.

Among prominent places of the Holy Land mentioned in the Old and New testaments are Bethany, Bethel, Bethesda, Siloam, Galilee, Jordan, and Ebenezer. All of these are reflected in Canadian place names, especially in Ontario. The places where Jesus was born and raised are not commemorated in English-speaking Canada, but both occur in Québec: Bethléem is 50 kilometres east of Sherbrooke, and Nazareth is a neighbourhood of Rimouski.

The community of New Sarepta, 40 kilometres southeast of Edmonton, was named in 1905 after Sarepta, now known as As Sarafand in present-day Lebanon. Sardis, a community in the city of Chilliwack, B.C., was named in 1888 for the site of one of the seven churches of Asia, in the western part of present-day Turkey. There is a Troy 25 kilometres west of Hamilton, Ont., another 25 kilometres east of Chatham, Ont., and yet another 10 kilometres northwest of Port Hawkesbury, N.S.

The Plain of Sharon is a fertile region on the Mediterranean coast north of present-day Tel Aviv. The reference in the Song of Solomon ('I am the rose of Sharon ...') inspired David Willson in 1831 to name the Children of Peace community of Sharon, 5 kilometres north of Newmarket, Ont. At the north end of the Plain of Sharon is Caesarea. It is also the name of a community 25 kilometres southwest of Lindsay, Ont. Its name is of dual origin, since the place was settled by the Caesar family in 1836.

The biblical stories and classic tales of the Middle East and the Holy Land reinforced the resilience and perseverance of Canada's nineteenth-century pioneers. Today, the names add variety to the toponymic fabric of Canada, especially in Ontario, New Brunswick, and Nova Scotia.

Capitalizing on Foreign Capitals

The map of Canada shows a sprinkling of place names derived from world capitals, reflecting the strong attachment many settlers felt towards their homelands. Most, predictably, are named for capitals in northern Europe and the British Isles. A national capital possesses a measure of prestige and an aura of strength and longevity, but the history of these Canadian namesakes is not always based on such stalwart qualities.

For example, the town of Paris, Ont., 10 kilometres west of Brantford, was not named for the grand French city of *savoir-vivre*. Rather, this Paris derived its name from the development, in 1829, of local gypsum deposits that were used to make plaster of Paris and fertilizer.

The community of Warsaw, Ont., with a population of 230, situated 20 kilometres northeast of Peterborough, was named by Thomas Choate in 1841 after his birthplace, Warsaw, N.Y., which was itself named for the Polish capital.

One might conclude that the village of Hague, 50 kilometres north of Saskatoon, was named for the Netherlands' seat of government, The Hague, but it was really named for J.H. Hague, a railway engineer who later ran a hotel in Vancouver.

The majority of capital-city namesakes in Canada, however, were legitimately named for foreign capitals. Athens, Ont., an attractive village of 960 people, 20 kilometres west of Brockville, was named in 1888 to honour the great city of Greece and its respect for education. The village, which then went by the mundane name of Farmersville, had three schools, including a teacher's training school.

The city of London, Ont., on the Thames River, takes its name from the surrounding London Township, itself named for Great Britain's capital in 1798. In 1791, Lt.-Gov. John Graves Simcoe wanted to locate the capital of Upper Canada at the Thames River site, then known as The Forks. He called it Georgina, for the reigning monarch. York (now Toronto) was subsequently chosen as the capital of Upper Canada and, in 1825, Georgina and The Forks were replaced by London when a post office was opened and the seat of government for London District was moved there. It was moved from Vittoria, a small community near the town of Simcoe, 70 kilometres east of London.

A few communities in Canada were named for capital cities that were prominent in the news. In the early 1870s, Belgium made headlines when Britain secured its neutrality during the Franco-Prussian War. In 1871, Brussels, the anglicized version of Belgium's capital, was chosen to replace the community name Ainleyville and the postal name Dingle, 85 kilometres north of London.

Belgium's capital also turns up in its French form, 85 kilometres southeast of Brandon, Man. A settlement of Belgians, Bruxelles was named in 1892 by Catholic archbishop Alexandre Taché.

In 1854, the Crimean War and Russia in general were popular topics of discussion. In that year Springfield, Ont., changed its name to Moscow, commemorating the retreat of Napoleon from the Russian capital in 1812. Moscow, with a population of 65, is located 30 kilometres northwest of Kingston.

But capitals in the news were not always looked upon favourably. During the First World War, tensions arose over the loyalty of the predominantly German residents of Berlin, Ont. The name had been chosen for a post office in 1833 by the early settlers. Berlin was settled first by Mennonite families from Pennsylvania in 1807, later by Pennsylvania 'Dutch' families (of German stock) and, after 1830, by German immigrants. In 1916, amid much rancour, a vote was taken to change the name. Among six choices, Kitchener was the most popular. It commemorates Lord Kitchener, the hero of Khartoum who drowned that year.

Khartoum, the capital of Sudan, also spawned another Canadian place name. Khartum, Ont., a locality in Renfrew County, was named in 1909 to commemorate Ottawa Valley rivermen who joined the Nile expedition in the 1880s. Two road signs identify the limits of the place, but there was only a single family living in it in 1999.

A Canadian traveller can arrange an extensive tour of European capital cities without having to take along a passport. Amsterdam is 55 kilometres north of Yorkton, Sask. It was homesteaded about 1912 by settlers from the Netherlands, and was named for that country's administrative centre.

Stockholm is a village of 390 people, 60 kilometres south of Yorkton. The colony of New Stockholm was settled by Swedes in the 1880s. When the Canadian Pacific Railway arrived there in 1903, the name Stockholm was proposed for a station and post office.

The community of Copenhagen, with a population of 200, is 20 kilometres southeast of St. Thomas, Ont. Recalling the capital of Denmark, it had a post office from 1870 to 1914. Vienna, a former village of 480 people (now part of the Municipal Township of Bayham), 35 kilometres southeast of St. Thomas, was named in 1836 by Capt. Samuel Edison (the inventor's grandfather) for the Austrian city where one of his ancestors was born.

Dublin is the name of a community of 255 people, 30 kilometres northwest of Stratford, Ont. Named Carronbrook in 1854, it was renamed Dublin in 1878 to honour settlers from Ireland. New Dublin is a small community of 60 people, 14 kilometres northwest of Brockville, Ont. Dublin Shore and West Dublin are seaside communities 18 kilometres southeast of Bridgewater, N.S. An Irish settlement was located in that part of Lunenburg County in the 1760s and was named New Dublin Township.

Other modified capital namesakes include New London, a rural community 38 kilometres northwest of Charlottetown, P.E.I. New London was established as a port in 1773 on Grenville Bay, which became known as New London Bay by the mid-1830s. A former farming community 25 kilometres northeast of Edmonton had a post office called New Lunnon from 1876 to 1907. This spelling reflects a Cockney pronunciation of London. The neighbouring town of Gibbons has preserved the name in Lunnon Drive.

Cairo, Ont., Bagdad, N.B., and Damascus, Ont., are Canadian communities representing Middle East capitals. Manilla, a community of 450 people, 20 kilometres west of Lindsay, Ont., replaced Mariposa in 1857 as the postal name. It is named for Manila, the capital of the Philippines. Delhi, 15 kilometres west of Simcoe, was laid out in 1828 as Fredericksburg, but when that name was found to be in use elsewhere, the township name Middleton was given to the post office in 1831. The name was changed to Delhi (pronounced 'DEL-HYE') in 1853, in honour of the then-capital of India, but the name Fredericksburg remained in general use even after the second change.

Settlers may have had great expectations for their new towns, hoping they would one day deserve their proud, capital names. But with the exception of London and Paris, none of the namesakes has reached a population of more than 1,000. Some, such as Reykjavik, Man., Riga, Sask., and New Kiew, Alta., are now nothing more than wind-blown fields. Then again, these founders of capital communities may not have been attempting to build great cities; perhaps they were simply seeking a fond reminder of their former homes or a curious footnote in place-name geography.

Anzac to Zealandia: Names from Down Under

In another essay, I referred to Canadian place names found in New South Wales, Australia, including *Toronto* and *Canada Bay*. That prompted several readers from Down Under to send me more examples: *Labrador* in Queensland; *Canadian Bay* and *Sherbrooke* in Victoria; and *Quebec, Alberta, Calgary, Toronto, Vancouver, Ottawa,* and *Ontario* avenues in Clapham, near Adelaide, South Australia. Their interest has inspired me to pursue this southern link further, by researching names from the South Seas that are found in Canada.

Before his luckless third voyage to Canada's Arctic, Sir John Franklin served as governor of Van Diemen's Land – later renamed Tasmania – from 1836 to 1843. Franklin disappeared in 1845 along with his entire expedition to the Northwest Passage. At the request of Franklin's wife, Lady Jane, a group of islands on the west side of the Boothia Peninsula was named the Tasmania Group in 1859, in honour of Sir John's service in Tasmania. In 1910, they were designated Tasmania Islands.

Lighter emotions inspired other names. In 1858, a group of settlers set sail from Charlottetown, P.E.I., to begin a new life in New Zealand. At the same time, another group from Charlottetown moved east to the area of Souris, P.E.I., to take up farming. They called their settlement New Zealand – a light-hearted reference to the first group's destination – and it later became the name of their post office from 1879 to 1914.

In 1871, residents needed a name for the new post office at Davis Corners, west of Perth, Ont. The postal inspector proposed Zealand, in honour of the South Seas country, shortening the name for postal efficiency. Although the office was closed in 1915, the rural community is still known as Zealand.

New railways frequently generated new place names. In 1905, the Canadian Northern Railway picked the name Ranfurly for a village 145 kilometres east of Edmonton. It was named for the Fifth Earl of Ranfurly, who had just completed a four-year term as governor of New Zealand. But CNR officials were thwarted when they chose the name Brock for a station on another line, 95 kilometres southwest of Saskatoon. The new settlers insisted on choosing their own name, and from several submissions they picked Thomas Englebrecht's suggestion of Zealandia. Englebrecht had emigrated from New Zealand on the S.S. *Zealandia* around 1907, and became one of the area's first settlers from abroad.

Australian Hugh McRoberts established Richmond Farm south of Vancouver in 1861. One of his daughters suggested that name after a favourite place in Australia (possibly the town of Richmond, near Sydney), according to Philip and Helen Akrigg in their *British Columbia Place Names* (1986). Richmond was subsequently chosen for the district municipality embracing Lulu Island, which is now a city of 125,000.

During the First World War, the Australian and New Zealand Army Corps (ANZAC) fought with distinction at Gallipoli, Turkey, and on the Western Front in Belgium and France. During that war, a tributary of the Parsnip River, north of Prince George, B.C., was named Anzac River in their honour. When the British Columbia Railway was built in 1967 from Prince George to Dawson Creek, a sixty-nine-car siding was built near the mouth of the river and called Anzac. Also during the First World War, land surveyor Frank C. Swannell undertook extensive surveys in the Nechako and Bulkley valleys west of Prince George. He named Bennett Lake for an Australian assistant who had served overseas with the Anzac forces. When informed that the province already had a Bennett Lake, Swannell recommended the subsequent change to Anzac Lake. The name Anzac was also given to a station on the Canadian Pacific Railway line just south of Fort McMurray, Alta., in 1922.

The Second World War created new place-name links with the South Pacific. Commonwealth Lake in northern Manitoba was named to recognize the British Commonwealth Air Training Plan of 1939–45. The islands in the lake were called Beaver for Canada, Kiwi for New Zealand, Kangaroo for Australia, and Lion for Britain. These appellations were given in 1974 by Allen Roberts, a New Zealand-born pilot who trained in Manitoba and subsequently became the province's director of surveys.

Sir Robert Richard Torrens was an Adelaide lawyer and state legislator who introduced the Torrens system of land titles in South Australia in 1858. In 1922, surveyor Richard W. Cautley named Mount Torrens, near Grande Cache, Alta., for him – sixteen years after Alberta adopted the Torrens system of transferring land ownership by registration and certificate, rather than by deed.

Australian prospectors travelled to the Yukon in the 1890s to probe the northern gravel bars for gold, even before the Klondike Gold Rush of the late 1890s: they brought a wealth of place names from down under. There are now two streams in the Yukon named Ballarat Creek, which recall a gold rush at Ballarat, in central Victoria, in the 1850s.

A tributary of Indian River, south of Dawson City, was named Australia Creek sometime between 1891 and

1894. A small tributary of that creek, called Melba Creek, is likely named after Dame Nellie Melba, a world-famous operatic soprano at the turn of the century and a native of Melbourne.

A tributary of the Klondike River, east of Dawson City, was also named Australia Creek in 1897. To avoid confusion with the creek named earlier, geologist W.H. Miller renamed it Aussie Creek in 1935. Australia Hill stands at the nearby confluence of the Klondike River and Hunker Creek.

Walhalla Creek, southeast of Dawson City, was named by a prospector who was believed to have worked the goldfields of Walhalla in Australia's state of Victoria, according to Robert C. Coutts in *Yukon: Places & Names* (1980). A few kilometres to the west is Australia Gulch.

Fiji may have been the birthplace of James Asasela, although Vilhjalmur Stefansson described him in *The Friendly Arctic* (1922) as a Samoan. 'Jim Fiji,' as he was commonly called, represented the Samoas in an exhibit of 'Native races' at the 1893 Chicago World's Fair. On the way home he boarded a ship in San Francisco that he assumed was sailing towards the Hawaiian Islands, because he saw some Hawaiians on board. But the ship was a whaling vessel that spent the next three years in the Arctic. He returned to the Arctic in 1896 for another three-year stint, and thereafter remained in the North. Asasela sailed with Stefansson's Canadian Arctic Expedition to the Beaufort Sea from 1913 to 1917. He then became a trapper, and disappeared about 1930 while sealing in Amundsen Gulf. In his memory, the Inuit of Paulatuk renamed Little Booth Island, east of Cape Bathurst, Fiji Island; and named an anchorage on the south side of the island Jim Fiji Harbour.

Motorists travelling between Winnipeg and Brandon pass signs for Sidney and Melbourne, only 8 kilometres apart, and may speculate about an Australian connection. In fact, no direct connection exists, although both Melbournes are named for the same British prime minister. Sidney is named for Sidney Austin, a British journalist who travelled on the first westbound CPR train (besides, the Australian city is spelled 'Sydney'). But there is still an impressive exchange of names that celebrate the enduring links between Canada and its Commonwealth partners in the South Seas.

A Touch of Portuguese on the East Coast

Portuguese navigators of the fifteenth and sixteenth centuries were tireless in their search for new lands, fishing grounds, and trading opportunities. Colonies flying the Portuguese flag sprang up around the world.

The Portuguese presence in Canada is reflected mainly on Newfoundland's east and south coasts, where such names as Bonavista, Burgeo, Fogo, and Ferryland are found. Labrador, too, is derived from Portuguese, and so is Bras d'Or Lake in Cape Breton Island.

From 1480 to 1509, Portuguese navigators accompanied several English expeditions from Bristol. João Fernandes, a *lavrador* ('small landholder') of Terceira in the Azores, may have sailed in 1497 to the New World with John Cabot. In 1499, he was directed by King Manuel I of Portugal to search for new lands. The following summer he sighted Greenland, which was soon identified on maps as Tiera del Lavrador. Some sixty years later, map makers restored the Norse name to that island and assigned the name Labrador to the mainland to the southwest.

King Manuel gave a similar directive in 1501 to Gaspar Corte-Real who, with three ships, explored the coast of

Labrador and the east coast of Newfoundland. The ship carrying Corte-Real disappeared while travelling separately, but the others returned to Portugal with information about the lands, the fishery, and the Aboriginal peoples. When Miguel Corte-Real sailed in search of his brother in 1502, he suffered the same fate.

The voyages of the Corte-Reals, and of subsequent Portuguese explorers, resulted in several names that are still in use today, either in obscure Portuguese forms or in English translation.

Cape Race, the most southeasterly point of Newfoundland, occurs on the earliest maps as *capo raso*, a description of the flat cliffs observed there. It also may have been named for Cabo Raso at the mouth of the River Tagus, the last point that sailors would see on leaving Lisbon.

Following the Newfoundland coast northward, one reaches Renews, a name identifiable with the Portuguese word *ronhoso* ('scabby'); rocks there are encrusted with shells and seaweed. Fermeuse and Aquaforte appear on a 1519 map as *R fermoso* ('beautiful') and *R da aguea* ('freshwater'). Ferryland may be traced to *farelhão*, meaning 'steep rock' or 'reef.'

Cape Spear, the most easterly point in North America, is shown on a map of 1505–8 as *Cauo de la spera* ('place of waiting'). Perhaps one of the vessels of the Corte-Real expeditions waited here for Gaspar or Miguel.

St. John's is believed by some to have been discovered by John Cabot on 24 June 1497, St. John the Baptist's Day. On a 1519 map, the entrance to the harbour is called *R de Sam joham*. Cape St. Francis appears as *c d s francisco*. The Portuguese identified Conception Bay with several variations of *baia da conceição*, a name that honours Our Lady of the Immaculate Conception. Baccalieu Island, off the tip of Bay de Verde Peninsula, is shown on a 1505 map as *y dos bacalhas* ('codfish island').

Some writers believe that Gaspar Corte-Real first applied the name Trinity Bay (in Portuguese, of course) to that body of water. While there is no cartographic evidence to support this, it is known that subsequent Portuguese travellers called it *baia de santa ciria.* The English seem to have chosen Trinity Bay in the late 1500s. Catalina, at the outer end of the bay, may have its roots in the name *p de S Catarina,* shown on maps of the mid-1500s.

Cape Bonavista is said to have been the first place reached by John Cabot in 1497. The name itself is Portuguese, appearing on early maps as *C de boa vista.* Cape Freels, from *y do frey luis* ('island of Brother Luis'), is the main landmark at the north side of Bonavista Bay.

Fogo Island and Cape Fogo, off the northeast coast, derive their names from the Portuguese *y do fogo* ('island of fire').

The first point west of Cape Race is Cape Pine, which occurs on sixteenth-century charts as *c de pena.* Across St. Mary's Bay is Cape St. Mary's. The bay is shown on a 1537 map as *Sa maria,* and the cape, on a 1536 map, as *Cabo de Sancta Maria.*

The French islands of Saint-Pierre and Miquelon were known in the 1500s as *Arcepelleguo das honze mill virgeens,* named for St. Ursula and the 11,000 virgins murdered by the Huns in the fourth or fifth century. A modified form of *virgeens* appears to have been transferred over 100 kilometres to the northwest to the present town of Burgeo and to Burgeo Islands.

Other south-coast names with probable Portuguese roots include: Great Colinet Island – *colmat* (possibly *colina* – 'hill'); Red Island – *ilha roxa*; St. Lawrence – *baia de sa lourenço*; Lamaline – *le belim* (possibly *baleia* – 'whale'); Cape Chapeau Rouge – *c roxo*; Fortune Bay – *fortuna*; Penguin Islands – *isla de pitogoen*; and Cape Ray along with several adjacent features with the specific, Codroy – *c rei* (and the Basque *Cap d'Array*). Cape St. George on Newfoundland's west coast may have begun as *c de S Jorge*.

João Alvares Fagundes, who sailed along the south coasts of Newfoundland and Nova Scotia about 1521, may have named Bras d'Or Lake to reinforce control of Portuguese discoveries near the boundary of Spanish exploration, set by the 1494 Treaty of Tordesillas. Bras d'Or is generally regarded as having evolved from 'labrador.' None of the other names from Fagundes's travels, including Cap Fagunda for present Cape Breton, survives.

The Basque Legacy on Canada's East Coast

For more than a century, the Basques ruled the fishery around the coast of the Island of Newfoundland and in the Gulf of St. Lawrence. In the sixteenth and seventeenth centuries intrepid whalers and fishers sailed every spring from their ports in the Bay of Biscay – where France and Spain meet – to their North American whaling stations and bountiful cod-fishing grounds. Although they left behind only scant vestiges of their occupation, the Basques did bequeath an impressive legacy of place names in Eastern Canada: more than 300 names can be traced to Basque sources or activities.

Many of those place names are found around the shores of the Gulf of St. Lawrence. Selma Barkham, an acclaimed expert on sixteenth- and seventeenth-century Basque activities on Canada's east coast, has documented the establishment of their fishing stations on the west coast of the Island of Newfoundland and the south coast of Labrador. She reported in 1989 that the Basques extended their fishing operations after 1590 to the Îles de la Madeleine and the western shores of the Gulf of St. Lawrence, including the area of the town of Gaspé.

Gaspé, where Jacques Cartier erected a cross and claimed the land for the king of France, has long been thought to be a name of Mi'kmaq origin. Basque writer Miren Egaña Goya challenged that conclusion in 1992, and claimed that Gaspé was derived from the Basque word *kaizpe*, suggesting 'shelter.' She cited seventeenth-century Basque maps by Denis de Rotis and Pierres Detcheverry, and the pilot books of Martin de Hoyarsabal (1579) and Detcheverry (1677). But not everyone agrees with her findings: Selma Barkham, for one, has supported the conclusion that Gaspé has Mi'kmaq roots.

Gaspé first occurred on Dutch maps in the 1590s, in the form of *Gaſpei*. During the next century it appeared in many forms on maps and in books: *gaspi, Gachepé, Gachepay, Gaschepay, Gaspay,* and finally *Gaspé* in French colonizer Nicolas Denys's *Description ... de l'Amérique septentrionale*, published in 1672.

In 1609, historian Marc Lescarbot listed *Gachepé* among words of Native origin. Until recently, it was widely accepted that Gaspé was derived from the Mi'kmaq word *gespeg*, meaning 'end of the point of land,' a suggestion advanced in 1833 by surveyor Joseph Hamel. This was confirmed by Father Pacifique, who worked among the Mi'kmaq at Restigouche, Qué., in the early years of the twentieth century, and who was described in 1928 by William F. Ganong, a New Brunswick natural scientist, as the best living authority on the Mi'kmaq. Ganong observed that the Mi'kmaq often adopted foreign names of prominent places, but he concluded that there was ample evidence Gaspé was from the Mi'kmaq language, expressing the idea of the 'last place of a territory.'

Ganong also considered 'Aspy,' in the name Aspy Bay, near Cape North on the east side of Cape Breton Island, to have the same origin as Gaspé. Selma Barkham has recently endorsed this conclusion. However, in a 1985 study of fishermen's contributions to early east-coast cartography and toponymy, she had concluded the name meant 'beneath the headland' in Basque, but did draw attention to *Isle de Gaspey* for Cape Breton Island on seventeenth-century maps.

One of the early European names for the Island of Newfoundland may have had a Basque connection. On many sixteenth-century Portuguese maps, the name Tierra de Bacalaos occurred in the area of the present island. A seventeenth-century French writer, using the form *Isles de Bacaleo,* attributed the name to the Basques. A 1710 French document, from the Basque town of Saint-Jean-de-Luz, claimed Basque fishermen were the first to discover Newfoundland in the fifteenth century, and to call it *Bacallao,* meaning 'cod.' Barkham has contended that it is more likely the Basques borrowed the word from the Portuguese. Ultimately, the English name Newfoundland supplanted the Portuguese/Basque name about 1502. Nevertheless, the small Baccalieu Island, off the north end of the Avalon Peninsula, two small islands called Bacalhao – one near Twillingate, and the other outside Labrador's Hamilton Inlet – and Baccaro Point in southwestern Nova Scotia are reminders of Portuguese/ Basque codfishing on Canada's east coast.

Near the southwest point of Newfoundland is the well-known ferry terminal of Port aux Basques, now part of the town of Channel–Port aux Basques. The Portuguese called the southwest point itself *Cabo do Rei* ('king's point'). When Captain James Cook surveyed the west coast in 1767, he rendered the Portuguese name as Cape Ray. Cook recorded the Basque variants *Cap d'Array* and *Cadarrai* as Codroy, today reflected in several local names, including Grand Codroy River.

Farther up the coast, the name Port au Port Bay is derived from the Basque *Ophorportu*, explained by Barkham as 'port of relaxation' or 'holiday port,' perhaps suggesting calm waters in the area enclosed by the huge hook of Port au Port Peninsula.

Continuing almost 250 kilometres to the northwest, the bays and harbours were either unprotected or too deep for the small Basque fishing craft to anchor safely. The next surviving Basque name is Ingornachoix Bay, from the Basque *Aingura Charra*, meaning 'bad anchorage.' North of the bay is the town of Port au Choix, on a narrow isthmus joining Point Riche Peninsula to the mainland. Selma Barkham reported that, when the fishermen learned how to place buoys, a small harbour with a narrow entrance on the outside of the peninsula became *Portuchoa*, 'the little port,' while the present Back Arm inside the peninsula became *Portuchoa Çaharra*, 'old little port.' Cook adopted the French adaptation of the name in 1767.

New Ferolle Peninsula, 40 kilometres farther north, was named *Ferrolgo Amuixco Punta* by the Basques, probably for El Ferrol in Spain's Galicia, where the whalers spent the winter. It was called 'new' to distinguish it from the original harbour of Ferolle, called *Ferrol Çaharra*, at present-day Plum Point, north of the peninsula.

In *The Canadian Encyclopedia* René Bélanger suggested a Basque origin for Mingan, on Québec's North Shore north of Île d'Anticosti. The Commission de toponymie du Québec, however, has concluded it is more likely a Breton name, meaning 'rounded stone.' On the eastern side of nearby Île Nue de Mingan, remains of ovens used by the Basques to render whale blubber account for the name Les Fourneaux, made official by the commission in

1986. Other ovens may have been situated on La Grande Basque and La Petite Basque, two of the seven islands at Sept-Îles. Baie du Havre aux Basques is at the south end of the Îles de la Madeleine. Anse aux Basques is a cove near Chandler, 55 kilometres south of Gaspé.

Basque fishing and whaling activities were especially intense on the St. Lawrence where it is joined by the Saguenay. Ten kilometres upriver is Anse du Chafaud aux Basques, where Samuel de Champlain reported in 1612 he found Basque fishermen drying fish on their *échafauds* ('scaffolds' or 'flakes'). Twenty-five kilometres downriver from the mouth of the Saguenay is another cove called Anse aux Basques.

Opposite the mouth of the Saguenay, and near Trois-Pistoles on the east shore of the St. Lawrence, is Île aux Basques. Several ovens on the island have been abandoned for some 350 years, but the locality name Les Fourneaux remains on current maps. In 1981, the government of Québec created the regional county municipality of Les Basques, with an area of 1,130 square kilometres and a population of some 10,500. Trois-Pistoles is its main urban centre. In 1996 the federal electoral district of Kamouraska–Rivière-du-Loup became Kamouraska–Rivière-du-Loup–Les Basques.

Several other names may have Basque roots, including: Scatarie Island, offshore from Cape Breton Island, from the Basque *escatari*, 'stacks,' reflected in a the offshore Cormorandière Rocks, some as high as more than 6 metres; Placentia, in Newfoundland, transplanted from the port of that name (now Plencia) in the Basque country of Spain, and confirmed, among others, by E.R. Seary in *Place Names of the Avalon Peninsula* (1971); Burin Peninsula, also in Newfoundland, from the Basque word *burua*, 'head'; Cap Dégrat, at the north end of Newfoundland's Northern Peninsula, from the Basque word *grat*, meaning a place where codfish is prepared; more than 140 places – including Barachois-de-Malbaie, near Gaspé – incorporating the generic term 'barachois' or 'barasway,' derived from the Basque *barra-txo-a,* meaning 'small bar'; and finally, more than 100 names incorporating the Basque-derived word *orignal,* used by the French to identify the North American moose.

Spanish Names along Our West Coast

In 1596, an old sea captain called Juan de Fuca met Michael Lok, an English trader, in Venice, and told a fantastic tale of exploration and discovery along North America's northwest coast four years earlier. The mariner explained that he had been instructed by the Viceroy of Mexico to seek the fabled Straits of Anian, a water body believed to provide a convenient, practical sea passage between Europe and the Far East.

Nearly thirty years later, Lok's tale was recounted in *Purchas His Pilgrimes,* a world history by Rev. Samuel Purchas. Lok reported that the sea captain had claimed to have discovered, between the 47th and 48th parallels, a wide inlet leading into a broader sea with many islands, and that there was 'an exceeding high Pinacle, or spired Rocke, like a piller' at the entrance to the strait.

Some 150 years later, in 1778, when Capt. James Cook explored the same coast, he missed the entrance because bad weather forced him to sail north far from shore. Nine years later, Capt. Charles W. Barkley, an English fur trader, sailed by the entrance. Knowing about Lok's story, he decided the name of the old seaman should be commemorated. The latter had been born Ioannis Phokas (not Apostolos Valerianos, as often reported) on the Greek Island of Kefallinía (Cephalonia), but in the service of the

Spanish for some forty years he bore the name of Juan de Fuca. A year after Barkley's trip, Capt. John Meares entered and charted the strait. He partially retained Barkley's designation, giving it the name John de Fuca.

Thus Barkley and Meares made the Lok tale seem authentic, although many historians and explorers thereafter wondered whether de Fuca had only woven a clever tale, and that it was simply a matter of luck that the area's geography partially fitted Lok's interpretation. In 1792, Capt. George Vancouver used Straits of Juan de Fuca on his chart, although he questioned the tale's authenticity. Research by historians in Spanish and Mexican archives has failed to reveal any evidence of Juan de Fuca's voyage.

Capt. John Walbran, who knew the British Columbia coast intimately and who studied its geographical names thoroughly at the turn of the twentieth century, agreed with Barkley's decision, declaring that de Fuca 'was undoubtedly the discoverer of the strait which bears his name.' Walbran went on to declare that a spire at the entrance to the strait next to Tatooche Island in present-day Washington state fitted the description by Lok, and thus it received the name Fuca Pillar.

Curiously, the Geographic Board of Canada delayed until 1934 its approval of the name of the water body which forms Canada's boundary south of Victoria and Vancouver Island, and then identified it as Juan de Fuca Strait. The United States board approved the style Strait of Juan de Fuca, so both forms occur on present-day maps and charts.

During his voyage of discovery, Capt. Vancouver became friends with the Spanish naval officer Juan Francisco de la Bodega y Quadra, and decided that the main island on the west coast should be called Quadra and Vancouver's Island. Ultimately the single name Vancouver Island came into use. In 1903, the Geographic Board of Canada named Quadra Island near Campbell River for the Spaniard.

There were several Spanish voyages along Canada's west coast from 1774 to 1792. Among the explorers honoured with names of geographical features are Gonzales López de Haro (Haro Strait), Alexandro Malaspina (Malaspina Strait), Jacinto Caamaño (Caamaño Sound), Dionisio Alcalá Galiano (Galiano Island), and Cayetano Valdés (Valdes Island). Cortes Island and Hernando Island recall the conqueror of Mexico, Hernando Cortés.

Other noted Spanish officers and administrators commemorated include Pedro Alberni (Alberni Inlet), Luis de Córdova (Cordova Bay), Manuel Antonio Flores (Flores Island), Juan Maria Lasqueti (Lasqueti Island), Felix de Texada (Texada Island), Vicente Tofino de San Miguel (Tofino Inlet), and Ciriaco Ceballos (Zeballos Inlet).

Many Spanish names have been superseded, among them Gran Canal de Nuestra Señora del Rosario la Marinera for the present Strait of Georgia, Río Florida Blanca for Fraser River, Entrada de Juan Pérez for Dixon Entrance, Puerto de Revillagigedo for Sooke Inlet, Rada de Valdés y Bazán for Royal Roads, and Bocas de Winthuysen for Nanaimo Harbour.

All told, Spanish exploration on Canada's west coast is recalled by about 100 names. These names, from Alberni to Juan de Fuca and Quadra, have provided us with a rich toponymic heritage.

Hispanic-American Names across the Land

Spanish explorers contributed to Canada's toponymic fabric on the Pacific coast, especially around Vancouver Island. Some of the more than 100 names of Spanish origin on the west coast recall late eighteenth-century exploration – from Juan de Fuca Strait to Port Alberni and Malaspina Strait – and are reviewed on pages 111–13. The community names Vittoria and Corunna and the township names Mariposa, Oso, and Zorra, among others, in Ontario recall battles between the British and the French in Spain during the Peninsular War, 1808–14. There are, however, a number of Canadian names that have been derived from Hispanic sources in the Western Hemisphere.

Situated on Ontario's Mississippi River, 45 kilometres southwest of Ottawa, Almonte (pronounced 'AL-MAWNT') is one of the province's prettiest urban settings in the 5,000-population range. Founded in 1819 by David Shepherd, Shepherd's Falls became Shipman's Mills in 1823 when Daniel Shipman bought the site, and built grist and saw mills. In 1837, its post office was called Ramsay, derived from Ramsay Township, and was located on the west side of the river. The east side developed as Victoriaville, named after the newly crowned queen. Ramsay and Victoriaville were united in 1853 as Waterford, which recalled the practice by men and horses of wading across the river above the falls. As Waterford was already the name of a place south of Hamilton, the postal inspector rejected the proposal. Two local residents, John Haskins and Col. J.D. Gemmill, suggested Almonte in 1856 for Juan Nepomucene Almonte, the Mexican ambassador to the United States. It appears they admired the ambassador for his fierce determination in promoting his country's rights at a time when there were widespread fears of American threats of expansion both to the north and to the south. On 1 January 1998, Almonte joined with the municipal townships of Pakenham and Ramsay to form the town of Mississippi Mills.

Juan Almonte was born in 1804, a son of Jose Maria Morelos, a leader of Mexico's independence movement in the early 1800s, who was executed in 1815 for insurrection. His mother was Brigida Almonte, an Aztec. Educated by a French teacher in New Orleans, Almonte rose to become a prominent military leader, fighting with distinction in several battles, including the Battle of the Alamo in Texas, in 1836. As well as serving as Mexican ambassador in Washington, he represented his country in London and Paris, where he died in 1869.

Spanish River, the longest river flowing from the north into Lake Huron, was first identified on a 1657 French map as *Aouechissaton*. Over the following 170 years it was identified as *R Tortue* ('turtle river'), *Estiaghicks*, and *R. Eskimanitigon*. In 1822, British Admiralty surveyor Henry Wolsey Bayfield was the first to refer to it as Spanish River.

Several theories have evolved to explain the origins of the name. One record suggests it was given by Bayfield in contrast to nearby French River. Dr. J.J. Bigsby, a geologist with the Canadian Boundary Commission, passed along the shores of Georgian Bay in 1823, and wrote in his *The Shoe and Canoe* that the area had been occupied by Spanish Indians. There is a claim that a Spaniard from the lower Mississippi during the old fur-trade days took refuge north of Lake Huron, and married an Ojibwa woman. Another tale claims an Ojibwa raiding party ventured far to the south, and returned with a Spanish woman. She married an Ojibwa chief, with the family taking on the surname L'Espagnol, with that evolving into Espaniel, a family name among the present-day Ojibwa. Another anecdote claims an Ojibwa chief, Spanish by birth, was captured by the Indians at a young age, and later married the daughter of a chief. When he became a chief he was assigned the area of the Spanish River as his hunting grounds, with the name recalling his roots. In the mid-1980s two Mexican coins minted in 1742 were found near the town of Massey, near the mouth of Spanish River, leading to speculation that they were lost by Spanish-speaking voyageurs.

The Spanish River Pulp and Paper Company built a mill in 1899, and called the place Espanola, meaning 'Spanish woman.' Purchased by the Abitibi Power and Paper in 1928, the mill was closed down during the Depression. Restructured as a prisoner-of-war camp in 1940, the mill was reopened in 1943 by the Kalamazoo Vegetable Parchment Paper Company, after the closing of the camp. Espanola was a company town until it was incorporated as a town in 1958. It now has a population of about 5,350.

The community of Toledo, Ont., 30 kilometres northeast of Brockville, had at least four other names before its post office was given this name in 1856. It may have been called after the Ohio city, but it may also have been named after the noted Spanish city.

More than one place in Canada was called California in the mid-1800s, when men heard the call to join in a gold rush while others set out to open new farming settlements; these were then labelled as though they might be sources of considerable fortune, but as likely were named in derision, as they failed to provide instant wealth. At least three settlements in New Brunswick were called California, but only the adjoining communities of Upper and Lower California, near Perth–Andover, remain as testaments to the search of fortunes during hard times. Two settlements in Eastern Ontario were also called California, but both have long ago been reclaimed by forests.

Two places in Saskatchewan have interesting origins with a Hispanic-American flavour. The village of Valparaiso, in the north–central part of the province, was named in 1904 when the Canadian Northern Railway was constructed from Portage la Prairie to Prince Albert. The name was proposed by postmaster George Green after the seaport city in Chile. That place had been named after a village in Spain. Situated only 10 kilometres from the towns of Tisdale, to the east, and Star City, to the west, Valparaiso never became a large urban centre, declining from 110 people in 1941 to only 30 in the 1990s. However, it continues to have a village council with a mayor and two councillors.

The town of Alameda in southeastern Saskatchewan, 55 kilometres east of Estevan, was named in 1883 by Christian Royer after the California city on the south side of Oakland. He had briefly visited California in the 1850s, but returned to Canada and took up land at the site of the village.

Spaniards Bay, Nfld., a town on the west side of Conception Bay, is located on a narrow cove called Spaniard's Bay. The bay's name was noted in 1610 by pioneer settler John Guy. It may have been given by English fishermen, who likely grouped Portuguese, Basques, and Spaniards under the single ethnonym. The town of Pasadena, Nfld., northeast of the city of Corner Brook, was named in the late 1930s after the city in southern California. This name is not, as might be presumed, derived from the Spanish language. It was devised in 1874 from the Chippewa (Ojibwa) language of the American Midwest, and means 'crown of the valley.'

Eldorado, Ont., the site of a gold rush in the late 1860s, is located in Hastings County, north of Belleville. Subsequently dolomite, hematite, marble, slate, and talc were mined in the area. Meaning 'the gilded one' in Spanish, the name refers to the mythical city of gold in South America. The community of Cordova Mines, northeast of Eldorado, was named in 1900, when a gold mine was opened there by the Cordova Mining Company. The company's name may have been given in honour of the Spanish city of Córdoba, noted for its fine gold- and silver-filigree ornaments.

Plumas, Man., is a small community near Neepawa. Originally named Richmond in 1879, it was renamed eight years later by postmaster James Anderson, who had arrived there in 1878 from Plumas County, California. In Latin the name means 'feather.' The area of the California county is renowned for its birds with beautiful plumage.

The famous Bonanza Creek, in the Yukon, was first known as Rabbit Creek. When gold was discovered in it in 1896 by George Carmack, Skookum Jim Mason and Tagish Charlie, Carmack proposed naming it after the Spanish word for rich ore deposits. Its largest tributary, Eldorado Creek, was named later that year by miner Knut Halstead.

Crimean Victories Commemorated across Canada

In 1854, in support of Turkey, Britain and France declared war on Russia. An expedition, under the allied command of Lord Raglan and General Saint-Arnaud, was mounted from Varna (in present-day Bulgaria) to destroy the Russian naval base at Sevastopol.

After landing on the west coast of the Crimean Peninsula on 14 September 1854, the allied forces engaged the Russians in a bloody battle at Alma, and moved on to Sevastopol, where they wintered after costly engagements at Balaklava and Inkerman.

The war continued through 1855, with the allies destroying Kinburn, the Russian base at the mouth of the

Dnieper, and the Russians taking Kars, in present-day eastern Turkey, after a heroic defence by the British general Fenwick Williams. In March 1856, Russia accepted the Treaty of Paris which, among other terms, gave formal recognition to the neutrality of the Black Sea.

The Crimean War has been commemorated in many place names across Canada.

News of the victory at Alma reached Canada in late October 1854 when the area of Rivière Saguenay near Lac Saint-Jean was being surveyed. The surveyors named an island in the river after the battle. The site ultimately became the city of Alma, which today has a population of 26,000. The territorial Alma Parish in New Brunswick was named in 1855 by a person who, on observing the site of the present village of Alma at the mouth of the Salmon River, was reminded of the heights above the Crimean battlefield of Alma. A community in Nova Scotia, another in Ontario, and possibly a third in Prince Edward Island were named Alma in honour of the victory in the Crimea.

The Battle of Balaklava was fought on 25 October 1854. The name Balaclava was given to communities in Renfrew and Grey counties in Ontario, and Balaklava was given to an island in Queen Charlotte Strait in British Columbia.

The Russian army was defeated at Inkerman on 5 November 1854. A territorial parish in New Brunswick, and a community in Dundas County, Ont., 40 kilometres southeast of Ottawa, were named for the victory.

The city of Sevastopol was finally captured in September 1855 after a siege of nearly a year. The name Sebastopol was given to a township in Renfrew County, Ont. Two defences of Sevastopol, Malakoff, and Redan, are recalled in the names of two small communities in Ontario, one in Carleton County, the other in Leeds

County. Malakoff, east of Moncton, N.B., was named by a surveyor to commemorate the battle.

Outside the Crimea itself, an engagement in 1855 at the mouth of the Dnieper River brought the name of the fortress called Kinburn into prominence. Kinburn, in Carleton County, Ont., became a post office at that time, and was named for the allied assault. Another Kinburn in Huron County, Ont., may have been named before 1855 for a place in Scotland.

The valiant defence of Kars by General Fenwick Williams occurred when the community of Wellington on the Rideau River applied for a post office. As the illustrious hero of Waterloo had been already commemorated elsewhere in Upper Canada, the name Kars was given to this little place 30 kilometres south of Ottawa. A territorial parish in New Brunswick and a township in Northern Ontario were also named Kars, and in 1888 Karsdale was given as a postal name in Nova Scotia.

General Williams, the great hero of Kars, was born in Annapolis Royal, N.S. He became commander-in-chief of British forces in North America in 1859, and from 1865 to 1867 was lieutenant-governor of Nova Scotia. He was honoured in Nova Scotia with the community names Port Williams in Kings County and Fenwick in Cumberland County, as well as Fenwick Township beside Kars Township in Northern Ontario.

Also in Ontario: Varna, a community in Huron County, south of Goderich, recalls the Black Sea port where the allied forces gathered in 1854 for the expedition to the Crimea; Odessa, a community in Lennox and Addington County, is a reminder of an April 1854 bombardment of the prominent Black Sea port; and Kertch, a small place in Lambton County between London and Sarnia, commemorates an 1855 battle at the eastern end of the Crimea.

Brudenell Township in Renfrew County, Ont., Cardigan Strait in the Arctic, and Cardigan Rocks in Queen Charlotte Strait, B.C., recall James Brudenell, Seventh Earl of Cardigan, the leader of the famous charge of the Light Brigade at Balaklava.

Raglan Township in Renfrew County, Ont.; the community of Raglan in the city of Oshawa, Ont.; Raglan Range on Melville Island, N.W.T.; and Raglan Point on Balaklava Island in British Columbia were all given for James Henry Fitzroy Somerset, Lord Raglan, who died of cholera at the siege of Sevastopol in 1855.

Two other points of Balaklava Island were given names after officers: Nolan Point for Captain Edward Louis Nolan, who carried the order for the charge of the Light Brigade and was killed at the Battle of Balaklava; and Scarlett Point for Sir James York Scarlett, who led the charge of the Heavy Brigade at Balaklava. Adjoining Balaklava Island are Lucan Islands, which recall George Charles Bingham, Third Earl of Lucan, a commander of cavalry in the Crimean War. Lucan, a village 25 kilometres northwest of London, Ont., is also believed to have been named for him.

Griffith Township, adjacent to Sebastopol Township, in Renfrew County, probably honours the leader of the Scots Greys at the Battle of Balaklava, Colonel Henry Darby Griffith.

Cathcart in Brant County, Ont., was named for Sir George Cathcart, who was killed at the Battle of Inkerman in 1854.

St. Arnaud Hills on Melville Island, Nun., recalls the commander-in-chief of the French army, Marshal Jacques A. Leroy de Saint-Arnaud, who became ill at the Battle of Balaklava and died at sea in 1854.

Canrobert, a Canadian Pacific railway point in Ange-Gardien, 16 kilometres southwest of Granby, Qué., and

Canrobert Hills on Melville Island, Nun., were named for General François Certain Canrobert, who had succeeded Marshal Saint-Arnaud at the Battle of Balaklava.

The commander of the French fleet during the Crimean War was Admiral Ferdinand Alphonse Hamelin. He was honoured with the name Mount Hamelin on Melville Island, Nun. Hamelin's second in command, Admiral Bruat, is remembered in Mount Bruat, also on Melville Island.

Many thousands of soldiers died in the Crimea from untended wounds and disease. Into the horrible carnage came the indomitable founder of modern nursing, Florence Nightingale, who earned a reputation for her profound dedication to caring for the wounded and to improving sanitary conditions in field hospitals.

The community of Florence in Lambton County, Ont., was named for the Italian city in 1856, having been earlier called Zone Mills and Victoria. It may be more than coincidence in the naming of this place that the heroic 'lady with the lamp' gave the name Florence prominence. The village of Florenceville, the 'frozen french fries capital' in New Brunswick, received its name from the Angel of Crimea in 1855, having been earlier called Buttermilk Creek. About 1865, two adjacent elevations in the Coast Mountains of British Columbia were named Mount Florence and Mount Nightingale. The community of Nightingale, between Calgary and Drumheller, Alta., was named in 1911 in honour of the illustrious nurse.

Numerous streets and avenues commemorate the victories and officers of the Crimean War, among them being Raglan Street, the main street in Renfrew, Ont., and Alma Street in Vancouver.

Surprisingly, the most prominent name of the war, Crimea, was not applied to a Canadian feature. Nor was

Yalta, the Crimean resort community made famous as the meeting place of the allied victors of the Second World War.

Ladysmith to Mafeking: Names from South Africa

At the turn of the twentieth century, the Anglo-Boer War became the first war to be fought in the daily press around the world, as well as across the veld and on the kopjes of South Africa. From Monday to Friday, Canada's daily newpapers devoted their front pages to the hard-fought victories and the bloody defeats of Britain's military forces in their struggle to conquer the smaller, but better-led Boer forces. Two, and sometimes three, pages of the Saturday editions reported extensively on the marches, the sieges, the hardships, and the horrors of war, and each story was accompanied by numerous photographs of Britain's gallant officers.

War correspondent Winston Churchill launched his momentous career in South Africa. He was often right in the thick of the fighting, and even preceded the divisions of Field Marshal Frederick Roberts in the capture of Pretoria on 5 June 1900. During that year, both his name and a church on a hillside 15 kilometres west of Charlottetown prompted postal authorities to accept Churchill as the name of a new post office. Although the office lasted for only thirteen years, the little community of some fifty people still remains.

The taking of Pretoria was barely heralded in the press. Other than a school district in Manitoba and a street or bridge in a few cities, Pretoria was not assigned to Canadian geographical features. In contrast, the 118-day siege of Ladysmith, from 30 October 1899 to 28 February 1900, was front and centre almost every day. The heroic stories filed from Ladysmith won considerable sympathy around the world, especially in the countries of the British Empire.

The little Natal village, 200 kilometres northwest of Durban, and on the road to Johannesburg, had been named in 1850 for the Spanish-born wife of Governor Sir Harry Smith. In 1812, Juana Maria de los Dolores de Léon had escaped to the British lines during a battle at Badajoz, met the then Captain Smith, and later that year, at the young age of fourteen, married him. During Smith's long military and civil career, she accompanied him to Asia, Africa, and North America.

When news of the relief of Ladysmith reached Canada on 1 March 1900, James Dunsmuir, a coal-mining operator at Oyster Harbour on Vancouver Island (he was later a lieutenant governor of the province), named the settlement Ladysmith, and the post office was renamed later that year. The town now has a population of some 5,000. Upper Thorne Centre post office in Quebec's Pontiac County was changed to Ladysmith on 1 March 1900. At present the rural community has a population of nearly 100. Another Ladysmith post office was opened on 1 April 1900, 20 kilometres south of Sarnia, Ont., but the office was closed in 1912, and the name subsequently dropped out of use.

In 1913, the Surveyor General of Ontario named Ladysmith Township, north of Dryden in northwestern Ontario. Five adjoining townships were given names for other prominent South African places and British officers in the news: Colenso, Mafeking, .Redvers, Buller, and Wauchope.

Colenso, southeast of Ladysmith, Natal, had been named in 1855 for Anglican bishop John William Colenso. In 1879, when Sir Garnet Wolseley defeated the Zulus near Colenso, a post office was opened with this name, 25 kilometres northeast of Owen Sound, Ont. It was closed in 1915, and the name is now unknown locally.

The siege of Mafeking lasted for 217 days, from 13 October 1899 to 16 May 1900. Its defence was managed brilliantly by Colonel Robert Baden-Powell. The place name was derived from a Tswana word meaning 'place of rocks,' and is now officially spelled Mafikeng. Mafeking post office was opened on 15 June 1900, 20 kilometres north of Goderich, Ont. The office was closed in 1915, and, although the name remains in official use, the census has not recorded anyone living there in recent years.

In the early years of the twentieth century, the Canadian Northern Railway extended its line from Swan River, Man., to Prince Albert, Sask., and named three of its stations in Manitoba Mafeking, Baden, and Powell. Mafeking is now a community of 275 people, but neither Baden nor Powell has more than a dozen homes.

Near Matachewan, in Ontario's Timiskaming District, Baden Township adjoins Powell Township. Baden-Powell Lake in Algonquin Park, was name by London, Ont., scouts in 1969, after Robert Baden-Powell, the former military leader and distinguished founder of the Boy Scout movement.

In 1906, the Canadian Northern Railway built another line west of Swan River, Man., to Norquay and Sturgis, Sask. A station 25 kilometres southwest of Swan River was called Durban, for the largest city in Natal (now officially called Natal-KwaZulu). The city had been named for Sir Benjamin D'Urban, who had served as governor and commander-in-chief of British territories in South Africa from 1834 to 1838. A decade later he was appointed commander of the imperial forces in Canada, and died in Montréal in 1849.

Natal, a community in the District Municipality of Sparwood, B.C., and at the west side of the Crowsnest

Pass, was given its name during the Anglo-Boer War by the Canadian Pacific Railway. Natal Township, in Ontario's Sudbury District, was named in 1913 by the province's surveyor general.

Sir Redvers Buller was a popular, but indecisive military commander during the war in South Africa. Redvers, a town in Saskatchewan, 105 kilometres northeast of Estevan, was named in 1901. It grew from 843 people in 1981 to 937 in 1991, in contrast with decreased populations by most Saskatchewan towns of its size during the same period. Wauchope is on the Canadian Pacific Railway, 15 kilometres west of Redvers. It was named for Gen. Andrew Wauchope, a Scot who commanded the Highland Brigade at the Battle of Magersfontein, where he was killed on 11 December 1899.

One of the most crushing defeats of the British by the Boer forces, led by Gen. Louis Botha, took place at Spion Kop (meaning 'lookout hill'), 15 kilometres south of Ladysmith. A hill at the mining village of Phoenix, B.C., between the cities of Grand Forks and Greenwood, was named Spion Kop. And Spionkop Ridge and Spionkop Creek are in southwestern Alberta.

Louis Botha served as prime minister of the Union of South Africa from 1910 to 1919. His wise and tolerant leadership in melding the British and the Boers was widely hailed. In 1909, the Canadian Pacific Railway gave the name Botha to a station, 13 kilometres east of Stettler, Alta. On the same line a station was named Veldt, from the Afrikaans word 'veld,' literally meaning field, but used in general to describe an open, grassy landscape. In 1916, Botha River in northwestern Alberta was called after the South African statesman.

Jan Christiaan Smuts, another Boer military leader, succeeded Botha as prime minister of South Africa, and subsequently became a distinguished statesman on the world stage. In 1918, Mount Smuts, in the Rocky Mountains, 90 kilometres southwest of Calgary, was named when Smuts was a member of the British war cabinet. When the Canadian Northern built a line in the 1920s from Saskatoon to Melfort, a station 50 kilometres northeast of Saskatoon was named Smuts.

Sir Alfred Milner was appointed British high commissioner in South Africa in 1897, and two years later his unwise decisions precipitated the events that led to the outbreak of war between the British and the Boers. In 1902, he negotiated the peace with the Boers, and returned to England three years later. Milner Township in Ontario's Timiskaming District is 35 kilometres south of Matachewan. The community of Milner in British Columbia's District Municipality of Langley was named for him. In Vancouver Island's Prince of Wales Range, northwest of Campbell River, the 1,310-metre high Mount Milner was named for Viscount Milner. In the same range, Mount Roberts was named after Field Marshal Frederick Roberts, who replaced Redvers Buller as the commander of the British forces in South Africa in 1899, and Mount Kitchener was given for Lord Herbert Kitchener, Roberts's chief of staff. Mount Roberts, near the 49th parallel west of Rossland, B.C., was also named for the field marshal.

Few persons are more prominent in the history of southern Africa than Cecil John Rhodes, who distinguished himself in the development of the Kimberley diamond mines and, as a parliamentarian and an imperialist, vigorously pursued the creation of a 'British dominion from the Cape to Cairo.' A township in Ontario's Sudbury District was named for him, and nearby townships were called after Roberts, Kitchener, and Botha. Mount Rhodes is in the Rocky Mountains adjacent to the Clemenceau Icefield in British Columbia. It is near

Mount Livingstone and Mount Stanley, named in 1927 for those illustrious explorers of the 'dark continent,' David Livingstone, a Scottish missionary and Henry Stanley, an American newspaper reporter.

Kimberley, B.C., a city of 6,530 in the Purcell Mountains, may have been named about 1897 by Col. William Ridpath, of Spokane, Wash. Presumably he had hoped that the mines bought the previous year by his mining syndicate would be as rich as those of South Africa's Kimberley.

Except for Ladysmith, Qué., South African names replicated in Canada were confined to Ontario and the four western provinces. In Atlantic Canada, the naming process of geographical features had been largely complete by 1900. In Quebec, there was little sympathy for the names of a faraway war involving the suppression of a minority racial group seeking to assert its independence from the British Crown. And in the North, where almost no South African names were assigned, the events of the Anglo-Boer War were probably unknown to the turn-of-the-century explorers until well after their front-page publicity.

On the Western Front: Names from First World War Battles

Few European names are more honoured in Canada than Vimy. And rare is the city, town, or village that does not have a First World War monument with Vimy proudly displayed among the French and Belgian names of places where fierce and bloody battles were fought and more than 60,000 Canadians and Newfoundlanders lost their lives. Across Canada, more than 125 mountain, lake, river, township, village, and street names are perpetual reminders of these places.

Near the France–Belgium border, 125 kilometres southwest of Brussels and 160 kilometres north of Paris, Canadian soldiers valiantly fought with the Western Allies against German forces. They took part in many battles during the four-year war, but it was their capture of Vimy Ridge in April 1917, with 3,598 Canadian lives lost, that hastened Canada's elevation to equal partnership among the nations of the world.

In Ontario, the military barracks at Kingston is named Vimy; a locality 60 kilometres east of Timmins is called Vimy Ridge; and Vimy Ridge Island is in Lake of Bays near Huntsville, Ont. As well, thirty-five towns and cities from St. John's, Nfld., to Duncan, B.C., have a street named Vimy.

Several townships in Québec recall other Western Front battle sites, including six near Val d'Or: Vimy, Denain, Ypres, Cambrai, Lens, and Festubert. Many lakes and rivers in the same part of the province also reflect the names of engagements and skirmishes: Yser, Douai, Arras, Loos, St-Pol, Flanders, Courcelette, Courtrai, Langemarck, Messines, Poperinghe, Farbus, Verdun, Somme, Givenchy, Bruges, Armentières, and Amiens.

At Canadian Forces Base Valcartier, west of Québec City, there are mountains named for Vimy, Mont Sorrel, and Cambrai, and a river named after the Battle of the Somme, where more than 24,000 Canadians lost their lives in the summer and fall of 1916. The military post office on the base is called Courcelette, honouring a victorious Canadian advance in September 1916.

Vimy is also the name of a hamlet 65 kilometres north of Edmonton, and there is an imposing crest named Vimy Ridge in Waterton Lakes National Park in southwestern Alberta. The crest's steep northern flank is called Vimy Peak.

In Wells Gray Provincial Park, north of Kamloops,

B.C., a few features also recall the sites of battles – Vimy Lakes, Vimy Ridge, and two creeks named Lys and Fleurbaix.

In the Canadian Rockies, Valenciennes River, Flanders Mountain, Mons Peak, and Verdun Glacier recall battles and actions on the Western Front.

During the First World War, the highlands 115 kilometres east of Edmundston, N.B., were opened for settlement. A French-speaking community there, which was called Anderson Siding until 1920, changed its name to St-Quentin to honour a battle that took place in 1918. St-Quentin is now a village of 2,300, serving a large farming and logging region.

In 1915, H.G. Richards of Saskatchewan was killed during the Battle of Hooge in Belgium. In the 1950s, his brother proposed that a new customs post 140 kilometres south of Swift Current be named Hooge. Officials liked the spirit of Richards's idea but instead of Hooge they chose the name Monchy, a place near Hooge where a decisive battle in the breaking of the Hindenberg Line took place in 1918.

Messines, Qué., 100 kilometres north of Ottawa, was named in 1921, replacing the name Burbridge, which had been given for the federal deputy minister of justice who had prosecuted Louis Riel in 1885. Messines, Belgium, was the site of important offensives in 1917 and 1918.

Armentières, as in the marching song 'Mademoiselle from Armentières,' is a French town on the Belgian border where a number of engagements took place. It appears twice on the British Columbia coast: as the name of a channel on the west side of the Queen Charlotte Islands and as the name of a rock on the west side of Vancouver Island, 45 kilometres southwest of Port Alberni.

Flanders, Belgium, the scene of many First World War battles, is the name of a township in Northern Ontario, a railway siding 255 kilometres west of Thunder Bay, and a locality 25 kilometres east of Sherbrooke, Qué. The French form, Flandre, occurs as a township in Pontiac County, Qué., and is near Artois, a township named after a region in France.

Several townships across Northern Ontario also derive their names from the battles and engagements that took place along the Western Front. Among them are Verdun, Langemarck, Marne, Mons, Somme, and St. Julien.

One name almost always seen on First World War monuments is Passchendaele, the site of some of the war's worst slaughters. In 1931, the Post Office Department gave the name to its Dominion No. 4 office on Cape Breton Island. In 1940, the office became Glace Bay Sub Office No. 3, but Passchendaele remains the name of a neighbourhood in Glace Bay, N.S.

Only one Western Front battle name appears north of the 60th parallel, and that is Somme Creek in the Yukon, about 260 kilometres northwest of Whitehorse.

Heavenly Places from Coast to Coast

Giving names suggesting perfection on earth, or greater beauty than actually exists, is as old as civilization. Witness the biblical naming of Canaan, the land of milk and honey; or the naming of Greenland by Eric the Red in the tenth century.

There are names from one end of Canada to the other that reveal this seeking for perfection, for a heaven on earth, for some sort of never-never land. Perhaps the people who assign such names hope that an ordinary, even ugly, place will seem more beautiful, and thus a better place to live. Others may give such names as an ironic twist, knowing that the place has few good qualities.

In about 1670, Nicolas Denys, the governor of Acadia, was delayed by inclement weather for a week at the mouth of a stream flowing into Northumberland Strait, 30 kilometres northeast of present-day Moncton. He did not consider his stay unpleasant and called the stream 'Rivière de Cocagne.' Cocagne River recalls the fabled 'Land of Cockagne,' which was described in a thirteenth-century English satire as a mythical land of good food and happy times.

As early as 1684, the French called a section of the Annapolis Valley 'Paradis terrestre.' Subsequently, when the English arrived in the 1760s, they gave the name Paradise Brook to a tributary of the Annapolis River. The adjoining communities of Paradise and West Paradise, 8 kilometres east of Bridgetown, have together about 575 people.

Although it is a region with extensive rocky landscapes and an often harsh climate, Newfoundland has more places called Paradise than any other province. The town of Paradise is 12 kilometres west of St. John's. There is another community with the same name about 50 kilometres north of Grand Falls–Windsor, and a third on the Labrador coast near Hopedale. Paradise River is a community about 200 kilometres east of Happy Valley–Goose Bay, Labrador. Thirteen physical features, such as Paradise Arm, Paradise Cove, and Paradise Sound, are similarly blessed.

Québec has more than fifty features named Paradis, but many may have been derived from personal names. Ontario has fifteen lakes, points, and other features named Paradise. The pleasantly named Cootes Paradise, a marsh in the Hamilton suburb of Dundas, was named in the late 1700s for Thomas Coote, a British officer and keen sportsman.

The village of Paradise Hill, with a population of 455,

is located 45 kilometres northeast of Lloydminster, Sask. This is the only 'paradise' in Saskatchewan; Manitoba has none, but Alberta has five, and British Columbia, twenty. In the Yukon, there is a Paradise Creek, named in 1903 for its attractive setting.

References to the biblical 'Garden of Eden' occur in several Canadian names. Garden of Eden and Eden Lake in Pictou County, N.S., were named in 1830 when the area was settled by William McDonald, nicknamed Adam. Mount Adam north of the lake was named for him. The contradictory Garden of Eden Barrens is located south of the lake.

Ontario has fifteen features with Eden, including Eden Mills, Edenvale, Glen Eden, and two places called Eden Grove. Manitoba has five communities with Eden, including Edenburg, 20 kilometres west of Emerson, and Eden, 15 kilometres north of Neepawa.

There are four places with Eden in Saskatchewan. However, one of them, Edenbridge, does not get its first element from the biblical garden, but from the Yiddish plural of 'Jew,' as the place was originally a Jewish settlement. Edenwold, a village 35 kilometres northeast of Regina, named in 1890 by Austrian settlers, combines the biblical Eden with the German wald meaning 'forest'; postal authorities misread the 'a.' Alberta has two features with Eden, a lake and a First Nations reserve.

British Columbia has seven names with Eden combinations. Eden Mountain on Vancouver Island is adjacent to mountains named for Cain and Abel. The village of Kaleden in the Okanagan Valley was named in 1909 by Walter Russell, selecting the first element from the Greek word meaning 'beautiful,' and the second from the biblical garden.

There are six features in the Northwest Territories with Eden, but only one, Eden Island, has any relation to the biblical garden, the others being personal names. The island in Frobisher Bay was named in 1960 in association with a hydrographic station called Eve located there.

In about 1783, a surveyor called Buffington drew up a survey plan near the mouth of the Magaguadavic River in present-day New Brunswick, and lots were granted. It was subsequently discovered that a large lake existed where many of the lots were shown on the plan. Either Capt. Peter Clinch or Lt.-Gov. Thomas Carleton is reported to have said that the possibility of acquiring the grants in the lake was comparable to that of achieving the perfection described in Sir Thomas More's *Utopia,* and therefore assigned it the name Lake Utopia.

Twelve kilometres west of Barrie, Ont., is the small community of Utopia, named in the 1840s for More's fictional island. Utopia Lake, 40 kilometres north of Vancouver, was the site of the Utopia copper claim filed in 1908. Perhaps it was named because the lake is the utmost source of Britannia Creek and is nestled in a valley almost perfectly round.

Heaven occurs in four names in Ontario, including Blue Heaven Lake and Little Heaven Island, but is not an element in names elsewhere in Canada. There are three features with 'ciel' in Québec. More common are those with Valhalla, the Norse palace where the souls of slain heroes feast. Ontario has four Valhallas, including two called Valhalla Lake. Two communities, one in Manitoba and another in Alberta, are known by the name of Valhalla. In British Columbia, the word occurs in six names, including Valhalla Ranges, northwest of Nelson. Several other names in the ranges reflect a Norse theme. Valhalla Mountain on Baffin Island is near several other mountains named in association with Norse mythology.

Nirvana, a Sanskrit word suggesting a perfect state, occurs in two names in Canada. Lac Nirvana is in

Québec's Parc de la Vérendrye. Nirvana Pass is in British Columbia's Pantheon Range, which has many features named for mythological gods.

Achieving harmony in any community is difficult, but Nova Scotia has three places with the word, including Harmony Mills in Queens County. Prince Edward Island has a place called New Harmony.

In 1901, a Finnish utopian community was established on Malcolm Island, between Vancouver Island and the mainland, and named Sointula (Finnish for 'harmony'). Although the colony was disbanded in 1905, today Sointula is a fishing village with a population of about 675.

And then there are the heavenly names given by real estate developers ...

Canadian Names around the World

From Abilene and Buckingham to Valparaiso and Zurich, Canada has been a great importer of place names. Little known, however, is Canada's export of place names to other countries. Some of them, such as Inglewood, had been previously brought from the Old World to Canada. But most names carried to other countries – mainly to the United States and Australia – are genuine 'home-made' Canadian names. Some, like Canada and Québec, have even found their way back to the British Isles and France.

On the Island of Bute in Scotland, Canada Hill overlooks the mouth of the Firth of Clyde. From this vantage point, islanders used to watch vessels carrying friends and relatives to 'exile' across the Atlantic.

On the north side of Lincoln's Inn Fields, adjoining the Royal Courts of justice in central London, is a footpath called Canada Walk. Canada Way in Hammersmith,

5 kilometres west of London's Marble Arch, is part of a 1930s housing development that includes Commonwealth Avenue and Australia Road.

New Canada is a major rail junction just southwest of Johannesburg, South Africa. When and why the name was chosen is unknown to the names authority in that country.

On the Paramatta River in Sydney, Australia, is Canada Bay. This is where a party of French Canadians was exiled in 1839 after the aborted rebellions of 1837–8 in Lower Canada.

In the United States, Little Canada is a northern suburb of St. Paul, Minn. Settled in the 1840s by Benjamin Gervais and other French Canadians from Lower Canada, Little Canada became a township in 1858, a village in 1953, and a city in 1974. Today the suburb has a population of 8,000.

The names of Canada's two largest provinces are well represented in Europe and the United States. Ontario, a city of 100,000 people, located 55 kilometres east of Los Angeles, was founded in 1882 by George and William B. Chaffey, who were born in Brockville, Ont. Their success in developing irrigation projects for the citrus industry in San Bernardino County led to an invitation in 1887 to undertake similar work in Australia.

Ontario, Ore. (population 9,000), is situated on the west side of the Snake River – the Columbia River's largest tributary – 65 kilometres northwest of Boise, Idaho. It was named in about 1883 by James W. Virtue for his native province. Although Virtue was one of the founders of the town, he actually settled 120 kilometres northwest of there in Baker, where he became the sheriff.

Québec is a popular designation for places and streets in France, England, the United States, and Mexico. Jean Poirier, the dean of Québec toponymy, has counted fifty-one 'Québec' streets and roads in France alone. In 1980, the city of Paris named Place du Québec in its Latin Quarter in honour of the close ties between France and Québec. In Acapulco, Mexico, the Plaza de Québec was inaugurated in 1983 as a sign of friendship between the citizens of Acapulco and the people of Québec.

Quebec is also a village 8 kilometres west of Durham in northern England. This authentic Canadian name first appeared as the name of a farm on a British topographical survey map in 1862. The mists of time have obscured the reason why the name was given.

Eighty kilometres southeast of Nashville, Tenn., is an unincorporated village called Quebeck. The unusual spelling was given in 1889 by J.S. Cooper, the town's founder, who constructed a sawmill there. The name was inspired by Cooper's travels in Québec, where he was impressed with the province's lumber industry.

Toronto may be the most common Canadian name transferred abroad. The best-known is Toronto, Ohio, a city of 7,000 on the Ohio River, 50 kilometres west of Pittsburgh. It was named in 1881 by Thomas M. Daniels in honour of the home city of W.F. Dunspaugh, the chief stockholder of the Great Western Fire Clay Company, then a principal industry in eastern Ohio.

In 1844, George W. Thorn, a native of Ontario's capital, founded the village of Toronto, Iowa, about 40 kilometres northwest of Davenport. There are a number of smaller American centres named Toronto, in Indiana, Kansas, Missouri, and Tennessee.

There is also a town called Toronto in New South Wales, Australia, 140 kilometres north of Sydney. According to the philatelic journal *The Canadian Connection*, it was named in 1888 to mark a visit that year by Ned Hanlan, the Ontario-born world sculling champion in 1880. The town is located on Lake Macquarie, Australia's largest permanent salt lake, which is noted for its water sports.

Montréal, Canada's second-largest metropolitan area, is the name of two small communities in Arkansas and Missouri. There is also a city of 900 called Montreal in Wisconsin; it is near the Montreal River, which forms part of the Wisconsin–Michigan boundary.

Incidentally, none of the places called Ottawa in the United States traces its origin back to Canada's capital. American cities such as Ottawa, Ill., and Ottawa, Kans., were named for the Ottawa (now known as the Odawa), who migrated from the area north of Lake Huron to the American Midwest.

Inglewood, Calif., is another Canadian export. Located 13 kilometres southwest of downtown Los Angeles, the city of 725,000 is the home of the Great Western Forum, where the National Hockey League's Los Angeles Kings team plays its home games. Inglewood owes its origin to a visit in 1887 by the sister-in-law of N.R. Vail, a developer. She lived in the little Ontario community and railway junction nestled in the Credit River valley, 20 kilometres northwest of Brampton. Inglewood, Ont., was named in 1885 by Thomas White, the local member of Parliament, after a place in England.

Except for a few street names in cities such as Atlanta, Orlando, and Los Angeles, it as difficult to find any of Western Canada's cities, towns, and villages repeated outside of Canada. An exception is Klondike, which occurs in several villages and townships throughout the United States. At least one book, *Michigan Place Names*, claims the name recalls the gold rush that took place in the Klondike region of Alaska!

As I was completing research for this essay in 1994, an acquaintance told me that an ancestor of his wife, a Simcoe, Ont., farmer and preacher named Abraham Austin, built a Baptist church in 1884 at a crossroads 75 kilometres north of Birmingham, Ala. First called Austin City, it was renamed Simcoe that same year. The little village of three churches and several houses still exists. I wonder how many other Canadian names are dotted across the United States and in other countries of the world?

Revealing Special Characteristics of Place Names and Generic Terms

Cities: Source of Pride, Broken Dreams

Teeming millions grow
And the anxious world rolls on;
Brilliant cities tall and wide
Boast their numbers and their pride.

–from the sonnet 'Liberty' by Archibald Lampman

For growing municipalities, the ultimate stamp of success is to achieve city status. Indeed, in the past many frontier communities added the title to their names to give the impression of stability and permanence.

The exception is the province of Québec. Since 1968, it has used the term *ville*, not *cité*, to designate urban centres, be they cities or towns. Montréal and Québec City (officially, Québec) were once known as *cités*, but there are now only two incorporated *cités* in the province: Côte-Saint-Luc and Dorval on the Island of Montréal.

Over the past eighty years, the number of incorporated cities in Canada has nearly doubled as a result of increasing urbanization. That number would be tripled if the 75 Québec *villes* with more than 15,000 people were included.

Most of Canada's 137 incorporated cities are in

Ontario and British Columbia, with 51 and 34 cities, respectively. In 1911, there were 74, with British Columbia leading the way with 25, followed by Ontario with 19 and Québec with 10.

The most populous incorporated city in 1911 was Montréal, with 470,500 inhabitants, followed by Toronto, with 376,500. Although Toronto now is referred to as Canada's largest city, it was technically in second place, with a population of 635,395, in 1994. The largest city in population then was Calgary, but on 1 January 1998, Toronto became a city of 2,385,421 by absorbing the four cities and one borough surrounding it.

The city with the smallest population – 725 people – is Greenwood, B.C., located near the United States border east of the Okanagan Valley. In 1911, it had 778 people, but it was not the least populated then. That year, the city of Sandon in the West Kootenay had only 151 people, down from about 5,000 when it was incorporated in 1898. Sandon is now a ghost 'city.'

Founded in 1785, Saint John, N.B., has the distinction of being Canada's oldest city.

Timmins, Ont., covers the largest area of any Canadian city, embracing several urban centres and vast expanses of lakes, rivers, and unpopulated woodland over 3,000 square kilometres. The smallest city, with only 2.79 square kilometres, had been Vanier, but it was amalgamated with Ottawa on 1 January 2001. The smallest city in the area is now Outremont, Qué., with 3.67 square kilommetres.

No place with 'city' in its name is incorporated as such. Labrador City, Nfld., a town of 9,061 people, is the largest of these so-called cities. Others, such as King City, Ont., and Crystal City, Man., have maintained the title in their names with great pride and vigilance. The community of King City, 30 kilometres north of Toronto, had long been known by that name on registered plans, but postal authorities decided in 1841 that King would be enough. The Post Office Department rejected a change to King City in 1932, even though the Geographic Board of Canada authorized the change. Ontario incorporated King City as a police village in 1934. Finally, after persistent requests, the postal authorities accepted the change in 1953.

In the south end of Welland, Ont., is a little neighbourhood that proudly calls itself Dain City. Located where the Canadian National tracks cross an earlier route of the Welland Canal, it was officially called Air Line Junction in 1880 and Welland Junction in 1910. But in 1980, Dain City – named after a family that lived there during its early development – was recognized by the Ontario Geographic Names Board.

Places with the word *city* are rare in the Atlantic provinces. The only one other than Labrador City is Forest City in southwestern New Brunswick, a small community adjacent to Forest City, Me.

In Western Canada, many settlements included 'city' in their names with the expectation they would one day grow to live up to that title. Unfortunately, in many cases those dreams remained unfulfilled.

Dominion City, on the Roseau River, 80 kilometres south of Winnipeg, was first called Roseau. But when it was confused with Rosseau, Ont., Dominion was proposed in 1878 for a new railway station on the line linking Winnipeg to St. Paul, Minn. During the following two years of boom times, 'city' was attached to the name. Today, Dominion City has fewer than 400 people.

On the Little Saskatchewan River, 30 kilometres north of Brandon, Man., is a town of 406 people called Rapid City. The town was named in 1878 when an assembly of settlers rejected Saskatchewan City as being too long, and unanimously passed a resolution to append 'city' to a rough English translation of Saskatchewan.

Crystal City is a village of 437 people on Crystal Creek, 110 kilometres southeast of Brandon. It was named about 1883 by its founder, Thomas Greenway, premier of Manitoba from 1888 to 1900. Greenway was optimistic it would become a great city. The village still proudly advertises itself as the 'Friendly City of the South.'

Located 100 kilometres southeast of Prince Albert, Sask., is Star City, a town of 507 people. It was founded by Walter Starkey, who homesteaded there in 1900. He chose the first syllable of his last name and added 'city' in the hope it would become a prominent place.

In 1954, after the discovery of uranium north of Lake Athabasca, Uranium City, Sask., became a town and municipal district. In 1973, E.T. Russell – author of *What's in a Name* – optimistically described it as having passed its boom stage to become a stable community of 1,000 residents. But its municipal offices were closed down in 1985 after Eldorado shut down its mining and milling operations.

Bow City, officially called Eyremore until 1958, was established on the north bank of the Bow River in 1924, 120 kilometres northwest of Medicine Hat. It never really developed beyond a few houses, with its population varying between 10 and 20 over the past thirty years.

The West Kootenay region of British Columbia has seen many instant cities come and, with few exceptions, disappear over the past 100 years. At the turn of the century, between the head of Kootenay Lake and Upper Arrow Lake were places called Duncan City, Trout Lake City, Circle City, and Dawson City. Near the northern tip of Lower Arrow Lake there were Burton City, Cariboo City, and Mineral City. With the exception of Burton, with a population of 130, little evidence remains of any of them.

Slocan, 30 kilometres northwest of Nelson, was an incorporated city from 1901 to 1958, when it was reduced to the status of a village. However, its Canadian Pacific station has retained the name Slocan City.

Queen Charlotte, the unofficial capital of the Queen Charlotte Islands, was grandly named Queen Charlotte City in the 1880s; this has remained a second official name for the community of 280 people.

The Yukon has also been the home of many instant cities. Among those that lasted long enough to be recorded were Calico City, Canyon City, Caribou City, Conrad City, Klondike City, Lynx City, Silver City, and Wind City. The most noted of all was Dawson City, which existed as an incorporated city from 1902 to 1904. It is now an incorporated town still widely known as Dawson City.

Most of these 'cities' sprouted overnight along rail lines or in bustling mining camps. Unfortunately, more often than not, the title did not work as a talisman to protect them from disappearing, if not during a single season, at least after a generation.

Sorting Out Ontario's Municipal Make-Up

Have you ever wondered what makes a city a city, or a town a town? On paper the definitions of each seem clear enough, but many exceptions exist. In Ontario, for example, population is used to define a city, a town, a village, and a municipal township. As we sort through the province's maze of municipal divisions, distinctions among them often become blurred.

While you might expect that most towns would have less population than places classified as cities, the *town* of Markham is found to have more than ten times as many people (165,000) as the *city* of Elliot Lake (11,600); the *township* of Oro–Medonte in Simcoe County also has more people (17,000) than Elliot Lake.

When the settlement of York became the city of Toronto in 1834, the new municipality had only 9,000 residents, growing to 30,000 by 1850. Since then, the minimum requirement for a city has been set at 15,000 residents for towns wishing to apply for city status, and at 25,000 for townships wanting to upgrade to the same level. The province at the end of 2000 had forty-eight cities, ranging from Toronto's 2,300,000 to three with fewer than the minimum requirement: Pembroke, Elliot Lake, and Dryden.

Pembroke had more than 16,500 people in 1971 when it became a city, but has since declined to fewer than 13,500. The former uranium mining community of Elliot Lake, between Sudbury and Sault Ste. Marie, has had four levels of incorporation since 1957. It was first created as a local improvement district by the province, with three appointed trustees to manage the urban infrastructure of the burgeoning community of nearly 25,000. After the decision by the United States in the early 1960s to drop uranium contracts with Canadian mining companies, the population plunged below 7,000. In 1966, Elliot Lake became a municipal township, with an elected council. Twelve years later, when it was incorporated as a town, its population had climbed to more than 8,000. Promoting itself as a retirement centre and tourist destination, the municipality grew to 15,000 when it was made a city in 1991. Since then its population has dipped to 11,600. Dryden had been a town until 1998, when it was allowed to upgrade to a city, although it had a population of only 7,731.

Perhaps the most unusual city was Nanticoke in the Regional Municipality of Haldimand–Norfolk. Covering a largely rural area of 653 square kilometres, and extending 28 kilometres inland from the Lake Erie shore, the city was created in 1974, with the prospect that a huge coal-fired generating station, a steel plant and a refinery would attract more industry and result in a population of some 130,000 in the city, and 300,000 in the region. However, an economic recession in the late 1970s and early 1980s curtailed growth, with the city population rising to only 22,000 in the late 1990s, and the region being held at 97,000. Legislation was passed in the Ontario legislature in December 1999 to disband the region, and amalgamate the city of Nanticoke, and the towns of Haldimand and Dunnville to form a new town of Haldimand on 1 January 2001.

Until the early 1970s, the traditional Ontario town was defined as an urban complex of 2,000 people or more, concentrated within small areas. Examples of such towns include Port Hope, Perth, Petrolia, and Penetanguishene, each with less than 12 square kilometres and fewer than 15,000 people. But at the end of 1999, Ontario had 18 towns with more than 250 square kilometres each, and 34 towns with more than 15,000 people.

With a population of 165,000, Markham is the most populous of Ontario's 117 towns. In 1994, I asked Mayor Frank Scarpitti if there were any intention of making it a city. 'While there will be debate whether it should be a town or city,' he told me, 'Markham will continue to convey the qualities of outstanding lifestyle amenities, low tax rates, and safe well-planned neighbourhoods, regardless what we call our community.' At the other end of the scale are twenty-one towns with fewer than the 2,000 required for town status, with Latchford (306) and Charlton (277) in Timiskaming District being the smallest.

The concept that the word 'town' should describe vast areas of mostly rural lands originated in a 1969 study of Muskoka District by Donald Paterson. He perceived an urgent need for integrated planning, managed development, and controlled pollution in some of Ontario's premier lake and cottage country. With the creation of the District Municipality of Muskoka in 1970, three towns were enlarged to embrace large areas of adjacent townships, with Gravenhurst expanding to nearly 490 square kilometres, Bracebridge to 620 square kilometres, and Huntsville to almost 690 square kilometres. The purpose of creating such towns was to design 'area municipalities' with uniform assessment over an area of dispersed permanent and summer communities, with a sufficient tax base to provide basic levels of municipal services, and to ensure better planning, regulated development, and controlled garbage and sewage disposal.

This new type of giant town was subsequently repeated in the regional municipalities developed after 1972. When the Regional Municipality of Sudbury was established in 1973, the new towns of Rayside–Balfour, Onaping Falls, Capreol, Walden, Valley East, and Nickel Centre were created outside the city of Sudbury. They ranged in size from Capreol, at 186 square kilometres, to Walden, with 769 square kilometres. Less than 5 per cent of the area of each of the regional municipality's towns was actually devoted to urban uses. On 1 January 2001 the region, the city, and the six towns were amalgamated to become the city of Greater Sudbury.

Gananoque, Prescott, St. Marys, and Smiths Falls are known as separated towns in the province, meaning that they do not participate in their respective county governments.

Have you ever heard of Clarington? With a population of 59,000 and an area of 576 square kilometres, it is in the Regional Municipality of Durham. Created as the town of Newcastle in 1974, it contained the well-known urban centre of Bowmanville. In 1991, after some fifteen years of public petitions, 59 per cent of the electors voted to change the name Newcastle. After a lengthy consultative process, a name-change committee proposed Clarington (an acronym formed from the names of the former townships of Clarke and Darlington). In order to retain provincial grants available to towns, the committee recommended keeping the status of a town municipality, but, because the electors expressed a dislike of the term 'town,' the committee proposed that Clarington be called simply a 'municipality.' Since then, five more 'municipalities' have been designated in the province, among them being Leamington and Tweed. (The 1999 *Ontario Municipal Directory* listed 21 'municipalities,' but its compilers had been misled by the awkward descriptions of them in the legislation.)

A village in Ontario is defined as an urban municipality with a minimum of 500 people and a maximum of 2,000. There were 19 villages with a population exceeding the maximum limit in 1994, but in five years that number was reduced to 11. Among the province's 116 incorporated villages in 1994, there were 16 with fewer than 500 people. By 1 January 2000, the number of incorporated villages was cut down to 45 as a result of mergers. Many of those will likely disappear into larger municipalities in the next few years.

There were 472 incorporated township municipalities in the province in 1994, but the number had been reduced five years later to 318 through amalgamation or redesignation. A municipal township is supposed to have 1,000 residents, but 127 had fewer than that in 1994, and that number was reduced to 84 on 1 January 2000. During the first ten years of the twenty-first century, most of these will be merged with adjacent municipalities, perhaps leaving a few remote townships in Northern Ontario.

In 1997 Ontario had 815 municipalities in total. Following instructions from the provincial government, mergers and amalgamations took place at a rapid rate across the province, with only 564 remaining at the beginning of 1999. Twelve months later the number had been further reduced to 517. By the beginning of 2001, the number dropped to 447.

Legacy of the Voyageurs on the Prairies

In the West, an isolated, often flat-topped elevation is called a butte, meaning 'small hill' in French. The word was likely first applied by French-speaking voyageurs, hunters, and trappers in the eighteenth century. It is one of several terms, such as prairie, meaning 'meadow,' and portage, meaning 'carrying place,' that were absorbed into the English language.

Butte (pronounced 'byoot') occurs in the names of sixty-six physical features and five communities in the four western provinces and the two western territories. Some are the most prominent feature for many kilometres around, examples being Lone Eagle Butte, north of Medicine Hat; Sarcee Butte, near Drumheller, Alta.; and Castle Butte, south of Regina. Nahanni Butte, adjacent to the Liard River in the Northwest Territories, is not a true flat-topped erosional remnant, but is called a butte because it is isolated from other hills.

Some of the names reveal fascinating tales of the West. Massacre Butte, near the head of the Crowsnest River in Alberta, was the place where twelve members of the Fiske Expedition were killed in 1867 by the Blood led by Medicine Calf. The town of Picture Butte, 12 kilometres north of Lethbridge, is named for petroglyphs on a former butte, which was excavated and is now the site of a subdivision.

Frenchman Butte, beside the North Saskatchewan River, northeast of Lloydminister, Sask., recalls the tale of a *canadien*, possibly a voyageur from Fort Vermilion, who was killed there by Aboriginals about 1800. Henry and Harry's Butte, near Shaunavon, Sask., is named after two half-brothers, Henry Thormeset and Harry Westvig, who homesteaded there.

The most common term for ravines and glacial spillways in the West is coulee (pronounced 'KOO-lee'), which occurs in 219 official names. The word comes from *couler*, meaning 'to flow' in French.

Coulees occur in areas where the topography has been extensively altered by glaciers, heavy rains, and melting snow. They are usually steep, extending back from a broad valley to the flat landscape above. A few, such as Chin Coulee and Etzikom Coulee in southern Alberta, are huge glacial troughs, often a kilometre or more wide, that cut across the short-grass prairie for some 50 kilometres.

Opinion is divided as to whether a coulee is a flowing water feature, a valley, or a combination of the two. Some descriptions in the toponymic records are contradictory. Philp Coulee, in Alberta, is described as 'a land feature [that] flows north from Montana into the Milk River.'

Coulees were best described by geologist Robert Bell in 1875: 'They are valleys or ravines with steep sides, often 100 feet or more in depth, which terminate or close sharply, often at both ends, forming a long trough-like depression; or one of the extremities of a coulee may open into the valley of a regular water course. The coulees sometimes run for miles and are either quite dry or hold ponds of bitter water, which evaporate in the summer and leave thin encrustations of snow-white alkaline salts.'

During the last quarter of the nineteenth century, the southern Prairie provinces were laid out in townships, ranges, and sections. The surveyors, accustomed to gullies and ravines back east, struggled with the unfamiliar term, often writing 'cullies' and 'coolies' in their reports.

Some coulee names have interesting origins. Etzikom Coulee takes its name from the Blackfoot expression for 'Crow Spring,' because the Crow war parties used to obtain water in that coulee. Chin Coulee is a translation of the Blackfoot *mistoamo* ('beard') because the slopes looked like a bearded chin from a distance. Blackstrap Coulee in central Saskatchewan recalls a tale of a pioneer settler of the Temperance Colonization Society in 1883, who spilled a crock of molasses on a bank of the coulee.

Although coulee occurs in names from the Red River Valley in Manitoba to the Peace River Valley in northeastern British Columbia, the densest concentration of named coulees is in the extensively eroded landscape surrounding the Cypress Hills of southwestern Saskatchewan and southeastern Alberta, and in the valleys of the South Saskatchewan, Oldman, and Belly rivers in southern Alberta. A marked characteristic of these hills and valleys is an annual water deficit for crop production. Various irrigation schemes have been implemented since the turn of the century to tap into the water collecting in the coulees, thus making them economically significant features in the landscape.

There are large areas of the Prairies with no named coulees at all, such as the Souris River Valley in southwestern Manitoba and southeastern Saskatchewan, and the North Saskatchewan River Valley of Alberta. Farmers and ranchers living there are likely to consider *coulee* a word out of a wild west novel, like gulch or canyon. The ravines and gullies in such areas usually contain flowing water during most of the year and, if they are given names, are usually called creeks.

All told, French-speaking explorers and map makers,

and later, English-speaking travellers and homesteaders, used words like coulee and butte for unfamiliar landforms, and thus contributed to the rich toponymic tapestry of the Canadian West.

Of Tickles and Rips, Barachois and Bogans

Giving names to land and water features is a fundamental activity of mankind, creating a meeting-place of culture with topography and hydrography. The Atlantic provinces have a fascinating variety of terms describing land and water features.

How many of us are aware that the generic term for a named feature, be it a *lake, creek, hill,* or *bay,* has been imprinted on our minds since early life so that we readily use the same term when we encounter similar features during our travels? Each of us is familiar with about 40 terms, and the kinds of features they apply to. However, English speakers in Canada use nearly 500 different terms in official names for Canadian land, water, and permanent ice features.

Some terms, however, are used for quite different kinds of features. Take gully, for example. In central Canada most people consider a *gully* to be a small ravine or a cut on a hillside formed by a heavy rain. But in the Maritime provinces, the term, derived from the French *goulet* ('neck'), is usually applied to narrow salt-water channels joining water bodies behind sand bars with the sea. The best-known ones in New Brunswick are Tabusintac Gully and Big Tracadie Gully. In Newfoundland, which has more than 250 officially recognized names with gully or gullies, the term is applied to three different kinds of features: a small pond with adjacent wetlands (Tom Howes Gully, Tilleys Gullies); a flowing watercourse

(Black Hummock Gully); and a coastal cove (Partridge Berry Gully).

Tickle occurs in nearly 250 approved names for water features along the coasts of Newfoundland and Labrador, and in 7 names in Nova Scotia. It is a narrow, treacherous salt-water channel, often characterized by hazardous tidal currents and rocks, both submerged and exposed. Among the many intriguing names in Newfoundland are Blind Mugford Tickle, Pinchgut Tickle, Headforemost Tickle, and Ingargarnekulluk Tickle.

Outside the Atlantic provinces, stretches of watercourses with no perceptible flow are usually unnamed in Canada. However, such features are frequently named in Newfoundland, Nova Scotia, and New Brunswick. The commonest word for a quiet stretch of water in Newfoundland is *steady,* approved in 71 names, including Elijaks Steady, Old Shop Steadies, and Rattling Brook Steadies. A similar feature in Nova Scotia is best known as a *stillwater,* where there are more than 175 with names like Black Rattle Stillwater, Corkum Stillwater, and Devils Funnel Stillwater. Nova Scotia also has 41 authorized names with *deadwater,* two being Coolan Deadwater and Barrio Deadwater. The most widely used term in New Brunswick for a smooth section of a river is *deadwater,* with nearly 60 names bearing the term. Pocomoonshine Deadwater, Pocowogamis Deadwater, and Semiwagan Deadwater are examples.

Barachois is used extensively for a coastal salt-water pond on the south and west coasts of Newfoundland, and along the Gulf of St. Lawrence shores of the three Maritime provinces and Québec. Originally applied by the Basques and the Acadian French to the gravel bar in front of the pond where fishing boats could be drawn up, the term was extended to the pond protected by the bar. Subsequently, it became an element in many other names.

In Newfoundland, the term *barachois* has not been standardized into a single spelling. The most frequent spelling there is *barasway*, as in Barasway de Cerf, Belleoram Barasway, L'Anse au Loup Barasway, Peltry Barasway, and 62 other approved names. As well, 29 names in Newfoundland use the form *barachois*, examples being Rocky Barachois Bight, Barachois Outside Pond, and Middle Barachois River. Just to be contrary, there is a locality on Hermitage Bay called Barachoix, with Barasway Harbour the name of the adjacent cove.

Nova Scotia has 34 *barachois* features, among them being Gabarus Barachois and MacLean Barachois. The term occurs eight times in New Brunswick and twice in Prince Edward Island. The community of Barachois-de-Malbaie on Québec's Péninsule de la Gaspésie has a population of more than 500.

A backwater channel adjoining a river in northern New Brunswick is commonly known as a *bogan*. The term, derived from the Mi'kmaq and Maliseet languages, is given to 30 names, including Fanton Bogan and Mersereau Bogan. A synonymous term of Mi'kmaq and Maliseet origin is *padou*, which occurs in two names upriver from Fredericton – Pickards Padou and Sugar Island Padou. Both terms occur in the name Banks of Padou Bogan on the Tobique River and, with a variant spelling, in Drakes Perdue Bogan on the Southwest Miramichi River.

A low-lying, seasonally flooded section along a river where wild hay can be cut is often called *intervale*. Nova Scotia has 16 named intervales (for example, Framboise Intervale) and New Brunswick has 12 (Meduxnekeag Intervale).

In Nova Scotia and New Brunswick, several flat, wet areas with low-growing vegetation are known as *heaths*. An example in Nova Scotia is Spinneys Heath; in New

Brunswick, Dorsey Heath. A similar feature is also called a *savannah* in Nova Scotia, where Black Georges Savannah and Dumpling Savannah are found.

A common term for rapids on rivers in southwestern New Brunswick is *rips.* Winding Stairs Rips and Meetinghouse Rips are two such rapids. Some turbulent waters in Passamaquoddy Bay also have named rips.

Aboiteau, a French word for a one-way sluice, has been extended to Parrsboro Aboiteau, a pond in Nova Scotia. *Drook* and *droke* are sometimes used to describe a grove of trees in Newfoundland, but in names they apply to brackish watercourses there and in Nova Scotia.

Among other unusual generic terms occurring in the names within one or more of the Atlantic provinces are: *gulch* (a ravine), *thoroughfare* (a channel), *soi* (a salt-water cove), *thrum* (an exposed rock in the sea), *nub* (a small island), *tolt* (an isolated hill), *vault* (a ravine), *mocauque* (a cranberry barren), *runround* (a subsidiary river channel), *flowage* (a shallow pond), *gut* (a narrow salt-water channel), *blow-me-down* (a steep headland), *brandies* (submerged rocks), *cay* (an exposed rock in the sea), *dyke* (a small watercourse), *lead* (a channel), *mash* and *mish* (a marsh), and *oven* (a coastal cave).

Shakespeare Remembered in Many Canadian Names

'I would to God thou and I knew where a commodity of good names were to be bought.'

– *Falstaff to Hal,* Henry IV, Part I, *I, 2, 92*

It was just a picture. But an evocative picture that produced the name of a Canadian city, and subsequently gave us a world-acclaimed dramatic festival.

As the story goes, Canada Company co-commissioner

Thomas Mercer Jones presented a portrait of William Shakespeare to William Sargint. It was hung in front of Sargint's inn at the point where the Huron Road, on its route from Guelph to Goderich, made a slight turn to the right to cross the Little Thames River, a tributary of Ontario's North Thames. Built in the spring of 1832, the Shakespeare Hotel was the first building in Little Thames. Within the next few months, the tributary became the Avon River, and the rising village was renamed Stratford on Avon – a name Jones may have been instructed to give it by the company's headquarters in London, England.

Stratford post office was opened in the fall of 1835, with this shorter form of the name eventually prevailing. It is interesting to note that its counterpart in England is called Stratford upon Avon, with the River Avon flowing not southeast into the River Thames, but southwest into the River Severn.

Although Stratford is widely known as the Canadian home of a Shakespearean festival, launched in the spring of 1953, the city is not overwhelmingly tarted up with references to the bard at every corner. Only 12 streets, out of a total of 239, have names associated with Shakespeare and his plays. A main north–south thoroughfare is Romeo Street, but there isn't one for Juliet. There is some consolation for her: a pontoon boat that treads the Avon's Lake Victoria is called *Juliet III*.

Behind the Perth County courthouse are the beautiful Shakespearean Gardens, designed in the early years of the twentieth century to reflect the plants and flowers mentioned in Shakespeare's works. In a city well known for its lovely green spaces, there is a park on the city's south side named for Anne Hathaway, the bard's wife.

All ten of the city's public schools have Shakespearean identities. As well as schools called Shakespeare and

Anne Hathaway, there are ones called Avon, Falstaff, King Lear, Hamlet, Romeo, Juliet, Portia, and Bedford.

Although some eighteen businesses in Stratford have capitalized on the word 'festival' to promote their outlets, almost none has utilized a Shakespearean connection to promote itself. Near the site of Sargint's original inn there is The Jester Arms, featuring English pub food, and on Romeo Street North is the invitingly named As You Like It Motel.

Thirteen kilometres east of Stratford is the community of Shakespeare, with a population of 600, settled in 1832 by David Bell. Its post office was Bell's Corners from 1849 to 1852, when the Post Office Department asked the residents to suggest a new name, to avoid confusion with another Bell's Corners in Carleton County. At a public meeting, Shakespeare was proposed, because it was quite near the head of the Avon River.

Over the years, several Shakespearean names were given in Northern Ontario. Shakespeare Island in Lake Nipigon was named in 1869 by geologist Robert Bell. In 1916, Louis V. Rorke, who was later to become Ontario's Surveyor General, named Shakespeare Township, 60 kilometres southwest of Sudbury, and Macbeth Township, 60 kilometres northeast of that city. In 1930, Rorke proposed an alternative for each of several common lake names in the area of Quetico Provincial Park. Among his selections from Shakespeare's plays were Mercutio, Modo, and Othello. Othello was not accepted, with the secretary of the Geographic Board of Canada declaring he saw no reason to change Wet Lake.

Hamlet Township, 100 kilometres northeast of Kapuskasing, was named in 1962. Hamlet was the name of a post office 10 kilometres southwest of Perth, Ont., from 1865 to 1879, when it was renamed Stanleyville. Twenty kilometres north of Orillia is the locality of Hamlet, named by the Post Office Department in 1932. On the closing of the office in 1955, the name was rescinded. After local protests, it was reinstated in 1978.

Shakespeare has been widely honoured in British Columbia, where its magnificent mountainous landscape motivated many to seek his muses for inspiration in naming mountains and railway stations.

In the 1920s, climber Neal Carter scaled a mountain 18 kilometres southeast of the present resort municipality of Whistler and named it Angelo Peak, likely for the naval officer in *Othello*. With a bit of whimsy, he made up the name Diavolo Peak to identify a mountain beside Angelo Peak, stating he had had a devilish time ascending it. In 1964, a team led by climber Karl Ricker explored the same area of Garibaldi Provincial Park, with the object of ascertaining high-level ski-touring routes. Finding the area sparsely named, Ricker formed a naming group, which included Dr. Carter. In commemoration of the 400th anniversary of Shakespeare's birth, they bestowed Mount Iago, from *Othello*, Mount Benvolio, from *Romeo and Juliet*, and Mount Macbeth.

In 1934, two adjacent mountains, 70 kilometres northwest of Campbell River, on Vancouver Island, were named for two of the greatest lovers of literature. Mount Romeo and Mount Juliet are separated by Montague Creek, taken from Romeo's family name, with Capulet Creek, from Juliet's family, being a tributary of Montague Creek.

In 1916, the Kettle Valley Railway built a line from Princeton, B.C., through the Tulameen and Coquihalla valleys to Hope, on the Fraser River. Seven stations in a stretch of 60 kilometres in the Coquihalla Valley were named for Shakespearean characters: Othello, Lear, Jessica, Portia, Iago, Romeo, and Juliet. After the Canadian Pacific closed the Coquihalla Valley section in 1962, the seven names were rescinded. However, Othello, only 5 kilometres from Hope, had developed into a small community, so its name was restored in 1985. The same year, the snow-avalanche section of the province's ministry of transportation and highways proposed the name Ophelia Creek for a small tributary of the Coquihalla River, 15 kilometres upstream from Hope. Juliet Creek is a tributary of the north-flowing Coldwater River, where the Juliet stop was located.

In 1960, climbers Robert West and Art Maki visited a large snowfield on the east side of Duncan Lake, some 50 kilometres north of Kaslo, on Kootenay Lake. Inspired by the name Duncan, West named the snowfield Macbeth Névé, and the breathtaking peaks surrounding the snowfield Mount Macbeth, Mount Banquo, and Mount Macduff. Subsequently, the snowfield was renamed Macbeth Icefield, and a fourth peak was called Mount Lady Macbeth.

In 1964, geographer David Harrison undertook a geomorphological study north of McBeth Fiord, on the east side of Baffin Island. Influenced by the fiord's name, given in 1945 for Royal Canadian Mounted Police Sgt. Hugh A. McBeth – who had led a patrol in the area in 1943 – Harrison proposed naming Cawdor Mountain, Thane River, Thane Lake, and Siward Glacier from Shakespeare's *Macbeth*, in honour of the 400th anniversary of the bard's birth. The names were rejected on the grounds that confusion might arise as to the origin of McBeth Fiord.

Falstaff Island, on the west coast of Hudson Bay near Rankin Inlet, is the only apparent Shakespearean name in the three northern territories. It was possibly named by geologist J. Burr Tyrrell, who, in company with his brother, James, explored the coast and adjacent river basins in 1893. Arctic explorers preferred descriptive names and the names of their crew, royalty, noblemen, financial supporters, colleagues, family members, and friends. Perhaps few travelled with copies of Shakespeare's plays.

There is a pair of Shakespearean names in Manitoba. Oberon, from *A Midsummer's Night Dream*, is a former railway station and post office, 20 kilometres south of Neepawa, and Portia, from *The Merchant of Venice*, is the site of a former school district, 100 kilometres to the northeast.

Near the head of the Ottawa River, 140 kilometres east of Val-d'Or, Qué., are two small lakes called Timon and Yorick, suggesting someone was familiar with two of the bard's plays. Two adjacent widenings of the Rivière Jacques-Cartier, midway between Québec City and Chicoutimi, are named Lac Obéron and Lac Titania, the king and queen of the fairies in *A Midsummer's Night Dream*.

In Nova Scotia's Guysborough County a place called Nerissa had a post office from 1900 to 1926, but the place has long been abandoned. The name likely came from the maidservant to Portia in *The Merchant of Venice*. It was rescinded in 1976, with Nerissa Round Lake remaining as a memento of the former community.

Christmas Creek ... and Other Yuletide Honours

Christmas is a joyous time, a season for celebration and glad tidings. It is not surprising, therefore, that many of the words and expressions associated with Christmas should be applied to geographical features across Canada.

In the remote forests of central New Brunswick a turbulent brook goes by the name of North Pole Stream. It was called that by the lumbermen in the mid-1800s, either because it was the farthest north that they had cut trees or because it was especially cold when they ventured towards its head.

In 1964, Arthur F. Wightman, then the New Brunswick member on the Canadian Permanent Committee on Geographical Names, gave the name North Pole Mountain to a 210-metre height at the head of the stream, stating that he was naming it for the mythical home of Santa Claus. To an adjoining peak he gave the name Mount St. Nicholas. This, in turn, inspired him to name eight nearby peaks for St. Nick's trusty reindeer. To the right of Mount St. Nicholas (looking downstream), he named Mount Dasher, Mount Dancer, Mount Vixen, Mount Prancer, and Mount Comet; to the left, Mount Blitzen, Mount Donder, and Mount Cupid.

Wightman derived these names from the famous poem by Clement Moore, *A Visit from St. Nicholas*, better known by its opening lines: ''Twas the night before Christmas, when all through the house ...' First published anonymously in 1823 in the Troy, N.Y., *Sentinel*, it was republished in the *Globe* of Toronto in 1854 under Moore's name. The version in the *Globe* refers to Dunder and Blixen, and turns Vixen into Nixen!

Continuing with this enchanting theme, Wightman proposed honouring the most famous reindeer of all, Rudolph of red-nosed fame. He was informed that such a name smacked of commercialism and did not have the sanction of long recognition in the literature of the Christmas season. So there is no Mount Rudolph in New Brunswick.

On the British Columbia–Alberta boundary, some 25 kilometres north of Kicking Horse Pass, is a towering mountain with the name St. Nicholas Peak. It was named in 1908 by A.O. Wheeler, a topographical surveyor, who was impressed with a resemblance to the profile of Santa Claus on one side of the mountain.

In the Yukon, two streams are called Christmas Creek. On the one that flows into Christmas Bay of Kluane Lake, Richard Fullerton and Rev. John Pringle, a Presbyterian minister, found and staked a claim on Christmas Day, 1903. On the other, a tributary of Matson Creek west of Dawson, gold was discovered on 25 December 1911.

British Columbia also has two streams named Christmas Creek, and one of them has Tiny Tim Lake near its headwaters. Christmas Hill and Christmas Point, and even Yule Rock and Yule Lake, are in the province too, but unfortunately none of the stories associated with them has been collected for the official records.

In Alberta, a tributary of the Athabasca River is called Christmas Creek, and Manitoba has three lakes called Christmas Lake. One of these is a translation of the Cree for Christmas, *Makosakeesekow.* Another was named in 1948 to honour Richard Edward Christmas, who was killed overseas during the Second World War while serving with the Royal Winnipeg Rifles.

On the Saskatchewan–Northwest Territories boundary is Santy Lake. It was not named for 'Santy' Claus, but for Flight Sgt. Samuel F. Santy of Moose Jaw, another casualty in the Second World War. Santa Lake in the Parry Sound district of Ontario was named for Pte. Ambrose Santa of Drayton, Ont., who was killed overseas in 1944.

In Ontario, there are two islands in lakes in Muskoka District Municipality called Christmas Island. Both were named in 1968: one had been given as a Christmas gift; the other is almost exactly halfway between the equator and the North Pole, home of Santa Claus. Santa's Village, a tourist attraction in Bracebridge, was named for the same reason. Another Christmas Island, near Kenora, was named in 1980 because the pine trees growing on it portrayed an image of the festive season.

Nova Scotia has two islands named Christmas as well,

both in Bras d'Or Lake, and may be named after a local Mi'kmaq family. The one near Eskasoni is known in Mi'kmaq as *Abadakwitcetc* ('a small portion set aside'). The other is 12 kilometres to the west; the community of Christmas Island is located nearby.

On the south coast of Newfoundland, at the mouth of La Poile Bay, is a prominent point called Christmas Head. Near the town of Grand Falls–Windsor is a body of water called Upper Christmas Pond, but the lake, formerly called Christmas Pond, upstream from Upper Christmas Pond, is officially Lewis Pond.

Québec has several features and places with the names Saint-Nicolas and Saint-Noël, with some of them no doubt associated with Christmas. Numerous lakes and other features in the province are called Noël. Near Joliette, northeast of Montréal, is the town of L'Épiphanie, which honours the Twelfth night of Christmas, sometimes called Little Christmas. The municipality of Biencourt, east of Rivière-du-Loup, was formerly called La Nativité.

And what is Christmas without the traditional Christmas feast?

Begin the celebration with Pointe à Champagne (Qué.) in Committee Punch Bowl (Alta., B.C.), with a reading from Charles Dickens Point (Nun.). Seat your guests around Dinner Place Creek (Man.). On each plate serve generous slices of roast Turkey Point (Ont.), accompanied by baked Potato Creek (Yukon), boiled Carrot River (Sask.), and Peas Brook (N.S.), and, of course, large dollops of Cranberry Point (P.E.I.). After the main course offer steamed Pudding Burn (B.C.), flambéd with Brandy Brook (N.B.), and a piece of pie made with Bakeapple Bay (Nfld.). Complete the dinner with Coffee Creek (Ont.) and selections from Lac des Bonbons (Qué.).

Season's greetings!

Of Valentines and Other Matters of the Heart

On 14 February 1718, a French missionary celebrated the first mass in a new settlement on the west side of Rivière Richelieu, some 20 kilometres south of present-day Saint-Jean-sur-Richelieu, Qué. In honour of the occasion, the settlers were given permission to name the place Saint-Valentin.

Valentine's Day embodies a pleasant mix of mystical romance, playfulness, and innocent amorous entreaty, usually expressed through florid, gushy cards, flowers, and chocolates. Why is there a joyous but somewhat frivolous day dedicated to a saint? Why Valentine? Why the middle of February?

Valentine's Day has its roots in the ancient Roman fertility festival of Lupercalia, observed on 15 February, when the first urgings of spring touched all the senses. Legend has it that one, perhaps two, Christians by the name of Valentine were martyred in Rome about 14 February in the third century. The early church adroitly associated the name of the saint, or saints, with the pagan celebration to sanction the festive practices of its Christian believers.

Besides Saint-Valentin, there are twenty-four features – mostly islands, lakes, and hills – in Canada with *valentine* or *valentin*. Many were named because of their distinctive heart shape, such as Valentine Lake in central Newfoundland and Valentine Lake near Hearst, Ont. An elevation with two pronounced summits overlooking Powell River, B.C., is known as Valentine Hill.

Some valentine names honour people. Valentine Creek, which flows north into Lake Diefenbaker in Saskatchewan, is named for George Valentine, an early settler. Valentine Lake, southwest of Miramichi, N.B., derived its name from A. Valentine Mitchell, a trapper who lived near it and adjacent Mitchell Lake at one time.

Among Newfoundland's many delightful names is Cupids, a town of 870 people located on the west side of Conception Bay, 30 kilometres due west of St. John's. John Guy of Bristol, who was appointed governor of Newfoundland in 1610, chose Cuper's Cove for the centre of a colony of thirty-nine settlers. Possibly because of the practice of suppressing the *r* by people in southern England, the forms Cupid's Cove and Cupids easily evolved, the latter form noted in 1630 by Nova Scotia founder William Alexander. Considering that Cupid was the Roman god of love, it is endearing to know that the wife of Nicholas Guie (not known to be related to the governor) gave birth at Cupids in the spring of 1613 to the first white child born in Newfoundland (see also pp. 216–19).

In Ontario, Cupid Island is in Georgian Bay's Thirty Thousand Islands, and Cupids Rock is in the St. Marys River southeast of Sault Ste. Marie. The Greek equivalent of Cupid is Eros, and this is reflected in the name Eros Lake, southeast of Geraldton, Ont. Beside it is Erato Lake, given for the Greek muse of lyric poetry.

In matters of the heart, Newfoundland stands out again as the province with the most appealing names. On the east side of Trinity Bay are Heart's Content, an incorporated town, noted by John Guy in 1612 as 'Hartes content'; Heart's Desire, also an incorporated town, possibly associated with a seventeenth-century vessel by the same name; and Heart's Delight, united in 1972 with nearby Islington to form a town with a population in 1991 of 880. On the west side of Trinity Bay is the community of Little Heart's Ease, with a population of 430.

The words *heart* and *coeur* occur in more than 100 names across Canada, often because of the shape of the feature. Noteworthy is Heart Shoal in Georgian Bay, whose outline on charts at the 10-fathom level is distinctly heart-shaped. Hearts Hill, a hamlet southwest of North Battleford, Sask., is adjacent to a range of hills with the configuration of a heart. Also in Saskatchewan, in Prince Albert National Park, is Hanging Heart Lake, whose northern part is shaped like a heart, although there is a legend that a heart was once found hanging from a tree. In Alberta, Heart Lake Indian Reserve is located beside a heart-shaped lake with a long point extending from its north side. Near Mount Waddington in British Columbia is Heartstone Peak, a mountain named for the large heart-shaped outcrop apparent on its south side.

At Motion Head in Newfoundland, a few kilometres 'up the shore' (actually south) from St. John's, is Bow and Arrow Shoal. If it could be aimed northwest, it would point at Merrymeeting Point, 4 kilometres away; or to the southwest for the same distance, at Hearts Point.

Except for Prince Edward Island and the Yukon, each of the provinces and territories has at least one feature with the words *love, lovers*, or *amour*. Both Newfoundland and Nova Scotia have little bays called Lovers Cove. Ontario has two narrow, secluded channels called Lovers Lane, one in the Thousand Islands of the St. Lawrence, the other in Georgian Bay's Thirty Thousand Islands. It also has a Lovers Isle on the west side of the Bruce Peninsula and a Lovers Creek near Barrie.

Two lakes with *love* in Manitoba are named for people: Love Lake near Neepawa for John Love, a homesteader; and Love Lake near Sherridon for Ernest Love, a mine official who was once chased by a bear when he was fishing there.

North of Nipawin, Sask., is a village called Love, named in 1930, possibly for an official of the Canadian

Pacific Railway, which passes through the community. There is another story of a lumberman becoming captivated by a young woman who cooked for logging crews at this rail siding, but it has the marks of a tall tale evolving from a name already given.

Québec has several features with *amour,* including Havre des Belles Amours, Ruisseau d'Amour, Baie des Amoureux, and Lac de l'Amour. In Labrador, adjacent to the Strait of Belle Isle, is a place called L'Anse-Amour but, alas, this is likely a misnomer, as the adjoining bay is called Anse aux Morts, perhaps recalling multiple drownings there.

A number of great lovers in history and legend are recalled in place names. On Vancouver Island, the first couple of the Creation are remembered in Adam River and Eve River. (On the current large-scale topographical map of the area, the compiler left the single river formed by the two unnamed, shrewdly leaving it to the map reader to discover which is tributary to the other!) Between the two rivers are prominent peaks called Mount Romeo and Mount Juliet.

In the Coast Mountains north of Vancouver, among a group of peaks named for classical figures, are Mount Eurydice and Mount Orpheus, separated by Styx Glacier. These names are drawn from the legend of the eternal separation of Orpheus from his great love, Eurydice, by the River Styx. Among the highest peaks of the northern Rockies in northeastern British Columbia are Mount Ulysses and Mount Penelope, named in 1961 and commemorating the Greek legend of Odysseus (Ulysses in Latin) and his faithful wife.

Now, for that favourite Valentine (Ont.) girl, send her a Bouquet Point (P.E.I.) of Flowers Cove (Nfld.) and a Gift Lake (Alta.) of a Box Canyon (B.C.) of Chocolate Cove (N.B.). She will send you back a pretty Card Lake (N.S.)

edged with Lacey Point (N.W.T.). After that you will go together to Dancing Hill (Man.) at Happyland Creek (Sask.), where you will exchange a Secret Creek (Yukon) and Lac Kiss (Qué.).

Giving the Devil His Due

Name givers and explorers have largely bypassed the Creator in Canada's place names. Gods Lake in Manitoba, Bay of Gods Mercy at the north end of Hudson Bay, and Islands of God's Mercie in Hudson Strait are rare exceptions. In contrast, the Devil and his associated names – Satan, Diablo, Mammon, Lucifer, Beelzebub, Sorcerer, and Windigo/Wendigo – have been widely honoured in Canadian place names. His traditional homes – Hell, Hades, and Enfer – are also prominently recorded from Newfoundland to the Yukon.

Newfoundland has thirty official 'devil' names, with several places called Devils Cove. Some of the more imaginative applications are Devils Dancing Place, a marsh on the Avalon Peninsula; Devils Dancing Table, a hill on the south coast; and Devils Dressing Place, another hill, this one beside White Bay on the north coast of the island. The province also has two places called Devils Kitchen, as well as a Devils Knob and a Devils Lookout Island. The French form *diable* occurs in Diable Bay and L'Anse au Diable Brook. 'Hell' turns up in Newfoundland names twelve times, including Backside of Hell Cove adjacent to Granby Island, Labrador; Hell Fire Pond on the Avalon Peninsula; and Hell Grapple Head on Black Island in the Bay of Exploits. Hells Al, a hill on the Avalon Peninsula, was found by Memorial University students in 1981; the name's significance was not determined. The Torngat Mountains in northern Labrador derive their name from the Inuktitut for 'home of the spirits.'

Prince Edward Island has only a single devilish name: Devils Punchbowl Provincial Park, at the site where more than 100 years ago John Hawkins lost a puncheon of rum in a deep hole and then recovered it.

There are forty names with devil in Nova Scotia. Among them are The Devils Burrow, a marsh near Windsor; Devils Cupboard, a cove on the Eastern Shore; Devils Funnel Stillwater, a pond near Liverpool; and Devils Limb, an island on the South Shore. Hell names are found in Hell Rackets, a reef in Mahone Bay; Hells Gates, a channel in Medway Harbour; and Helluva Hole, a pond near Windsor. Main-à-Dieu, in Cape Breton County, appears to mean 'hand of God,' but is really derived from a Mi'kmaq word for the Devil.

New Brunswick has eighteen devil names, examples being: Devil Pike Brook in Kings County; Devils Back Brook in Northumberland County; Devils Elbow Rapids in Restigouche County; Devils Half Acre, a hill in Albert County; another Devils Half Acre, a shoal in Passamaquoddy Bay; and Devils Hand, an 8-metre-high rock in Charlotte County, with huge finger grooves. Hellcat Brook, Hells Gate Rapids, and Hells Kitchen, a deep ravine in Fundy National Park, are examples of the province's eight hell names.

Québec has honoured the prince of demons in sixty-six place names, most of them in the French *diable.* Lac Batle-Diable, near Havre-Saint-Pierre, commemorates the achievements of one William Cormier, a local hunter and fisherman who was given the nickname which means 'beat the Devil.' Other *diable* names include Crique du Diable in Labelle County, Pointe au Diable in Soulanges County, Pont du Diable in Frontenac County, and several Rapides du Diable, Rivière du Diable, and Ruisseau du

Diable. Québec's twenty-three hell names are mostly with the French *enfer,* although there is a Lac Helluva near Schefferville. The province also has a Lac Hades, a Lac Lucifer, and a Mont du Sorcier. Windigo, the name for mythical cannibals of the Ojibwa and other Aboriginals, occurs in twenty names in Québec.

Ontario has given the Devil his due in seventy-eight places. My first encounter in the 1940s with a devil place name was the precipitous Devils Glen near Collingwood. I was told it was so named because a wrong step would surely lead to death. Then there is Devil Door Rapids, Devils Cellar Rapids, Devil Shoepack Falls, Devil's Warehouse Island, and Devils Punchbowl Lake. Hell occurs in seven names in Ontario, including Hellangone Lake in Parry Sound District and Helldiver Bay in Kenora District. Windigo, or Wendigo, is reflected in thirty-two names; Windigo Lake, Wendigo Lake, and Windigo Island each occur several times. Ontario also has three names with Hades, one being Hades Islands in Lake of the Woods. And there are four Styx names; River Styx is a treacherous area in the Cataraqui River section of the Rideau Canal system.

The Devil has nineteen names in Manitoba. Interesting ones are Daredevil Hill near Neepawa; Devils Punch Bowl, a crater-like ravine near Brandon; and Devils Portage near The Pas. Manitoba has two features called Windigo Lake as well as a Wendigo Beach, Wendigo River, and Wendigo Point.

Saskatchewan has recognized the Devil in eight names, among them Great Devil Rapids and Little Devil Rapids on the Churchill River. Helldiver Lake is near Cumberland House. Hellfire Creek, close to the U.S. border south of Gravelbourg, is said to have been named by a homesteader who found the area 'as hot as hell's fires.'

Alberta also has eight Devil names, including Devil's

Bite, a hill southwest of Calgary with one side gouged out, and Devils Thumb, a mountain near Lake Louise. Hell-Roaring Falls is a tributary of the Smoky River, south of Grande Prairie.

The rugged landscape and tumultuous rivers of British Columbia have evoked the Devil in twenty-nine place names. Among them are Devils Claw Mountain in Cassiar District, Devils Club Mountain in Cariboo District, Devils Den Lake in Alberni District, Devils Spire in Kootenay District, and Devils Thumb in Cassiar District. (Devil's Tongue, Devil's Tooth, and Devil's Rectum are unofficial names in the province.) In 1960, some mountains in an area of the Coast Mountains known as the 'White Inferno' were named for the mythical chief lords of Satan: Beelzebub, Azazel, Belial, Dagon, Moloch, and Rimmon. Mount Satan and Ogre Mountain also occur in the same group. North of the group is Purgatory Glacier and Styx Mountain. Also in the province are Lucifer Peak, Sorcerer Mountain, and Witch Spirit Lake. Hell appears in fifteen names in British Columbia, including the turbulent Hells Gate in the Fraser Canyon, Hell Gate in the Grand Canyon of the Liard, Hell Raving Creek in the Coast Mountains, Hellroarer Creek east of Okanagan Lake, and Hellsgate Slough beside the Skeena River. A set of dangerous rapids on the North Thompson River is called Porte d'Enfer Rapids.

In the Northwest Territories, Hell Roaring Creek is a tributary of the South Nahanni River, and Hell's Gate Rapids is a variant for Figure of Eight Rapids on the same river. Devil Island, in northern Nunavut, recalls the 1900 expedition of the Norwegian explorer Otto Sverdrup, which faced starvation on this island between Devon and Ellesmere islands. Hell Gate, in the same area, was also named by Sverdrup, who described the passage as a surging mixture of raging water. Tuurngatalik Island, on the south side of Baffin Island, gets its name from the Inuktitut expression 'it has devil spirits.'

The Yukon has three Devil names: Devilhole Creek, Devil's Club Creek, and Devils Elbow. Hellsgate Rapid is a perilous section of the Yukon River.

A computer search of the Canadian Geographical Names Data Base produced the lists of devil and hell names. The printouts also dumped out such personal names as Mandeville, Mitchell, and Burchell. When the first printout was made, smoke arose from the computer terminal, burning out a transistor!

Telenaming the Landscape

From Signal Hill in St. John's, Nfld., to Mount Marconi in southeastern British Columbia, and from Telegraph Narrows near Belleville, Ont., to Beacon Mountain in the Arctic Archipelago, Canada has a country-wide web of names relating to sending and receiving messages.

Signal Hill is probably Canada's best-known name associated with communicating over long distances. Its fame derives from experiments conducted there in 1901 by Guglielmo Marconi, who proved wireless transmission could be sent across the Atlantic. But the name has a much older history, reputedly dating from the late 1500s when a signal cannon was placed on the summit of the 152-metre height at the entrance of St. John's Harbour. Its earliest cartographic reference is James Cook's 1762 chart of the harbour. The hill, with its imposing Cabot Tower, was declared a national historic site in 1958.

In 1902, Marconi had four 65-metre towers constructed at Table Head in Glace Bay, N.S., and exchanged messages with Poldhu, Cornwall, England. The station was reconstructed south of the town in 1904, with the community there becoming known as Marconi Towers. The original site at Table Head is now the Marconi National Historic Site.

The inventor of radiotelegraphy has been honoured in other names across the country. A 63-metre hill with a radio tower in the Îles de la Madeleine is known as Butte du Marconi. Lac Marconi is east of Val-d'Or, Qué., and Marconi Township is north of Sudbury, Ont. Mount Marconi, in the Rocky Mountains, 60 kilometres east of Windermere, B.C., was named in 1919 after the 1909 recipient of the Nobel Prize for Physics.

Samuel Morse's invention in 1837 of the telegraph – the transmission of encoded messages through the opening and closing of electrical circuits – did not result in place names in Canada given directly for him, probably because it was too early in the process of commemorative naming in Canada to give an American inventor such an honour.

In the mid-1860s the Collins Overland Telegraph Company set out to connect New York with Paris by running a line across Bering Strait. More than 250 men were engaged to string the line through British Columbia, and more than $3 million were spent before the project was abandoned in 1866, when it was learned the Atlantic cable had been successfully laid under the ocean.

The surveys and subsequent construction of the Collins line provided several new place names in British Columbia and Yukon. At the mouth of the Skeena River, just south of Prince Rupert, large supplies of wire and other materials were landed in 1865, with the channel separating the mainland and Kennedy Island being named Telegraph Narrows that year. When the survey reached a widening of the Bulkley River, a tributary of the Skeena, it was named after Michael Byrnes, one of the explorers of the route. Burns Lake, as it was spelled, is considered the geographical centre of British Columbia, with the village of Burns Lake having a population of some 1,700. During a geological reconnaisance in 1875–6, George M. Dawson found the name Telegraph Range in use, 75 kilometres southwest of Prince George, with the Collins line having skirted its northeast side.

Although the Collins line did not get much beyond the Skeena River, its arrival had been anticipated when the surveyed route crossed the Stikine River in 1866 at an unnamed stream that became known as Telegraph Creek, with the small community of Telegraph Creek subsequently developing there. The planned route resulted in the names Telegraph Bay in Atlin Lake in northern British Columbia and Telegraph Mountain in Yukon, directly north of Lake Laberge. A member of one survey team was Michel Laberge, whose name was given in 1870 to the widening of the Yukon River, just north of Whitehorse.

Some 175 kilometres east of Edmonton is a small tributary of the North Saskatchewan River called Telegraph Creek. About 1885, poles were erected at its mouth in anticipation of a line crossing the North Saskatchewan. However, it was deemed too difficult to cross the river at that point, with the poles inspiring a name even though a line did not pass that way.

Telegraph Narrows in Ontario's Bay of Quinte commemorates a line that crossed from the mainland to Prince Edward County, often nicknamed Quinte's Isle. Within the narrows is the small Telegraph Island. Elsewhere in Canada there are Telegraph Coulee, 45 kilometres northwest of Saskatoon, Sask., Colline du Télégraphe, on the south shore of the St. Lawrence River, 60 kilometres northeast of Québec City, Île de la Ligne du Télégraphe on the north shore of the Gulf of St.

Lawrence, and Telegraph Hill, 15 kilometres north of Saint John, N.B.

The founder of the American Telegraph Company, which laid the cable under the Atlantic in 1866, was Cyrus West Field (1819–1892). In 1861–2, explorer Charles Francis Hall named Cyrus Field Bay, the first bay north of Frobisher Bay, on the east side of Baffin Island. When the Canadian Pacific was constructed through the Rocky Mountains in 1884, Field was there to observe progress. Field, the first station west of the divide, and nearby Mount Field were named for him.

The resort community of Dwight, 18 kilometres east of Huntsville, Ont., was named in 1871 after Harvey P. Dwight (1828–1912), then a Toronto-based official of the Montreal Telegraph Company, who enjoyed fishing in the area, and gave the place a telegraph office.

Telephone Lake in Morse Township (after prospector John Morse) northwest of Sudbury, Ont., with a line passing over a nearby hill, recalls the invention of Alexander Graham Bell. Telephone Bay in Opinicon Lake, north of Kingston, was named because a line was strung across the bay. Telephone Lake in western Manitoba was named when a line was strung for 105 kilometres over rugged Duck Mountain, with it only reaching the lake, two-thirds of the way, at the end of a season.

In 1951, a group of lakes north of Baddeck, N.S., were named Baddeck Lakes. A 1973 *National Geographic* map of the Atlantic provinces identified them as Grosvenor Lakes, after Bell's nephew, Gilbert H. Grosvenor, editor of the magazine from 1900 to 1954. In 1973, the local name was found to be Bell Lakes, although it was unclear whether they had been named for the inventor, who spent his summers at Baddeck. This name was confirmed by provincial authorities as being given in honour of him, and special commemorative maps were presented to museums and archives in Nova Scotia and Ontario.

In 1949, the Spruce Falls Power and Paper Company constructed a radio tower west of Timmins, Ont., and some 25 kilometres southeast of Foleyet. Three years later a radio tower was built by the Hydrographic Service on a small island south of Resolution Island, at the entrance to Hudson Strait. Although both towers were dismantled long ago, Radio Hill and Radio Island will remain as reminders of an important means of telecommunication a half-century ago.

Glaciologist Fritz Müller observed an imposing mountain on Axel Heiberg Island in 1966, and named it Beacon Mountain because he was reminded of coastal beacons used to guide ships past dangers and into harbours. The well-known Beacon Hill Park in Victoria, B.C., recalls two Hudson's Bay Company beacons placed about 1840 in such a way as to point towards the dangerous Brotchie Ledge.

A television repeater station was erected on a hill at Lumby, B.C., 23 kilometres east of Vernon in the early 1960s, with the site becoming known as Satellite Hill. Another kind of satellite was involved in the naming of the last piece of Canadian land discovered. In 1976, Betty Fleming, of the federal Department of Energy, Mines and Resources, was examining Landsat photos when she found a small 25-by-45-metre island, 33 kilometres southeast of Cape Chidley, Labrador's most northern point. When a Hydrographic Survey crew flew over it in a helicopter to confirm its existence, they observed a polar bear on it. Landsat Island was a natural name for it, and this was given to it in 1979. In 1978 the Soviet satellite *Cosmos* crashed in a lake near the Thelon River in the Northwest Territories. Cosmos Lake was given to it by Canadian military personnel, and confirmed in 1980.

From sixteenth-century fires and cannons through the years of wireless transmissions to the modern era's pulses from outer space, Canada's toponymy has reflected the development of communications. Perhaps future cyberian surfers will also contribute to the naming of landscape features.

A Giggle, er, Gaggle of Ha! Ha!s

The townsfolk of Saint-Louis-du-Ha! Ha! have long endured the amusement and disbelief of visitors, who are curious to learn how the place ever acquired such a comical name, and perhaps wonder which Louis among the saints was distinguished by his laughter. Perched by the side of a 380-metre hill, the village of 1,530 is just off the Trans-Canada Highway, 70 kilometres east of Rivière-du-Loup, Qué.

For many years, it has been speculated that the 'Ha! Ha!' part of the name recalled either a Huron word for 'portage' or an early explorer's exclamation. However, neither claim has any basis in fact.

The French word *haha* really means an unexpected barrier, or a dead end. Its use can be traced to the time of Joan of Arc in the fifteenth century. It was accepted by the French Academy in 1762 as a legitimate word to describe ditches outside openings in a wall around a park. The openings afforded a vista of the surrounding countryside while the hahas marked the park boundaries. On farms and in gardens in the south of England, a ha-ha is a sunken fence that segregates livestock from a manor house without interrupting the view across the estate's fields.

In Canada, the word *haha* first appeared on early maps of New France. A 1686 French map and eighteenth-century English maps show an island, a river, and a mountain called 'Haha' in the Bay of Fundy region.

So where is the unforeseen obstacle for which Saint-Louis-du-Ha! Ha! is named? Eight kilometres east of the village is 40-kilometre-long Lac Témiscouata. As early travellers headed north by canoe, they discovered that the lake curved to the northeast – away from their destination of the St. Lawrence River and Québec City. The little bay on the west side (mentioned in early records as Baie du Ha! Ha!, but never officially named) represented an unexpected and inconvenient dead end for travel by canoe. Travellers were forced to portage 80 kilometres west to Notre-Dame-du-Portage on the St. Lawrence. They walked through the community of Saint-Louis-du-Ha! Ha!, named after its founder, Louis Marquis. The municipal parish of Saint-Louis-du-Ha! Ha! was established in 1874.

But why the exclamation marks? It seems they were the result of another interpretation of the origin of the village's name disseminated by surveyors and writers, including Joseph Bouchette, the surveyor general and map maker of Lower Canada in the early 1800s; Bouchette's son, Joseph Bouchette, Jr., the deputy surveyor general; and Arthur Buies, a noted nineteenth-century historian and writer. They wrote that astonished explorers exclaimed 'ha! ha!' when they found their route was blocked, and subsequently named the body of water a 'Ha! Ha!'

A number of place names with the expression Ha! Ha! also dot Québec's Saguenay region. Eighty kilometres from the mouth of Rivière Saguenay, the waterway appears to divide into two major branches. The northern branch leads to Lac Saint-Jean, where the river rises. The southern branch is just as wide, but Baie des Ha! Ha! comes to a dead end within 10 kilometres. The plural form of the name is due to two coves at the head of the bay that act as hahas by giving the illusion of leading to another waterway. Flowing into one of the coves from the south is the 55-kilometre-long Rivière Ha! Ha! It drains two lakes called Lac Ha! Ha! and Petit lac Ha! Ha!

Baie des Ha! Ha! is also a prominent 15-kilometre inlet on the north shore of the Gulf of St. Lawrence. The bay gives the appearance of being the outlet of a major river, but in fact it leads only to several rocky barriers and small streams. The name Baye du Haha was in use as early as 1715. Ha Ha Rock is a 2-metre submerged hazard in Ontario's Georgian Bay.

Newfoundland has two widely separated inlets with 'Ha Ha' as part of their names. Ha Ha Bay is a 4-kilometre inlet near the north end of the Northern Peninsula, 15 kilometres west of L'Anse aux Meadows. A sand bar (or haha) at the south end of the bay, separating it from the appropriately named Isthmus Cove, prevents passage to Pistolet Bay to the south. On the east side of the bay stands Ha Ha Mountain, a 70-metre hill. Ha Ha Point is at the east entrance of the bay. The nearby community of Raleigh was originally called Ha Ha, but was renamed earlier this century by the Newfoundland Nomenclature Board. Several names of French origin in the area – Pistolet Bay, Falaise Point, Piton Point – suggest the bay was named by French fishermen or explorers.

The Ha Ha is an inlet at the town of Burgeo, on the south coast of Newfoundland. The head of this inlet is separated by a narrow isthmus (or haha) from another inlet called Long Reach. The Ha Ha was also likely named by the French during early exploration.

The proper spelling of place names such as Baie des Ha! Ha! has long been a source of debate among historians and map makers. In 1938, the Société Historique du Saguenay determined that 'haha' was the proper form, and encouraged historians to refer to the local inlet as Baie des Hahas. In 1947, Father Victor Tremblay wrote an extensive treatise on the origin of the word *haha* and concluded that the substantial historical and geographical evidence favoured the forms Baie du Haha and Baie des Hahas.

The Quebec Geographic Board (predecessor of the Commission de toponymie) informed the federal names office in 1947 that the correct form for the Saguenay inlet was Baie des Ha! Ha!, although acknowledging that 'haha' was an old French word meaning an unexpected barrier on a route. However, the Ottawa names office advised federal map and hydrographic chart makers that the correct name to use was Ha Ha Bay, the form authorized by Ottawa in 1901. It was only in 1960 that Baie des Ha! Ha! was approved for federal mapping.

Names with Ha! Ha! in Québec and Ha Ha in Newfoundland almost all identify water features that have land barriers. The present erroneous spellings have considerable historical precedence, so it would be most unlikely that the Commission de toponymie in Québec or the Geographical Names Board in Newfoundland would consider 'correcting' such names as Baie des Ha! Ha! to Baie des Hahas, or Ha Ha Bay to Haha Bay.

Saint-Louis-du-Ha! Ha! and The Ha Ha, among a dozen 'haha' place names, reflect more than 300 years of Canadian heritage – and quite unintentionally provide us with a little surprise and merriment.

Adopting Names of Native Origin and Acknowledging Names Used by Indigenous Peoples

Aboriginal Place Names: We All Have Our Favourites

No place names distinguish a country better than those of Aboriginal origin. And each of us likely has a favourite.

For some, it may be Antigonish, the small Nova Scotia town made famous for the cooperative movement which began there in the early 1930s. Reported to mean 'where the bears tear the branches off trees,' the name more likely means 'flowing through broken marsh,' in reference to the outlet of Rights River.

Québec is another favourite, meaning 'where the waters narrow,' in reference to the St. Lawrence upriver from Île d'Orléans. Its counterpart in Nova Scotia is Chebucto, the Mi'kmaq name for The Narrows in Halifax Harbour. The usual meaning given for Chebucto, 'a great long harbour,' is really just a variation of the idea of a narrow water body. Ontario, a pleasing name to many, is usually considered to be Iroquoian for 'handsome lake,' but may be better interpreted as deriving from the Huron for 'large lake.'

The southern British Columbia names Keremeos and Similkameen are deemed especially pleasing by some. Both are derived from Salish, the first meaning 'wind passage in the mountains,' the latter, 'swimming river.'

My own favourite is Kitwanga (pronounced 'KIT-wahn-gaw'), the name of a small community east of Prince Rupert, meaning, in Tsimshian, 'people of the place of many rabbits.'

No one can really deny the musical charm and sweet lilt of Athabasca, 'where there are reeds'; Madawaska, 'land of porcupines,' in New Brunswick; in Ontario, Madawaska literally translates as 'reeds at the forks of the river,' but the original name, Matouachita, points to Matouoüescarine, 'people of the shallows,' who were encountered by Champlain in 1613; Wascana, 'pile of bones,' the Cree name for Regina; Minnedosa, 'swift water' in Siouan, the equivalent of the Cree Saskatchewan; Winnipeg, 'murky water' in Cree, the equivalent of the Siouan Minnesota; Wawanesa, 'whippoorwill,' adapted from Wawonaissa in Longfellow's *Song of Hiawatha*; Coaticook, from Abenaki for 'river of pines,' a town south of Sherbrooke; and Muskoka, from Ojibwa Musquakie, their name for two chiefs given to Ontario's premier cottage country.

Some people, however, dismiss names of Aboriginal origin because of either their simplicity in meaning or the awkwardness of their pronunciation.

Some names that mean only 'big bay' (Malpeque), 'big water' (Meelpaeg Lake, Nfld.), 'big river' (Koksoak, Yukon), and 'fine river' (Oromocto) are occasionally ridiculed because of their ordinariness. Names that are difficult to pronounce are avoided by many – Musquodoboit, 'flowing out in foam'; Magaguadavic, 'big eels place'; Kashagawigamog, 'long and narrow'; Kaministiquia, 'place of islands in the river'; Illecillewaet, 'swift river'; Auyuittuq, 'place which does not melt' (the national park on Baffin Island); and Chinguacousy, 'young or little pines,' the township surrounding Brampton, Ont., where I was born.

Sometimes similar origins are revealed in names that do not appear to have common roots, examples being Timiskaming, Témiscouata, and Chicoutimi, with their reference to deep water; Baddeck and Petitcodiac, with their characteristic feature of bending sharply; and Saugeen, Mississauga, and Saguenay, with their reference to a river mouth.

Most spellings are only imperfect renderings of what Aboriginal people may have said to explorers, who then provided their interpretations to publishers and map makers. Names with the idea of 'ochre' were variously rendered as Romaine, Onaman, and Wunnammin, all from related languages. Chippewa and Ojibwa are different names for the same language. *The Handbook of Indians of Canada* (1912) lists 169 variants of Ottawa – Ahtawwah through Outaouais to Watawawiniwok.

The varied Aboriginal languages in Canada have given our toponymy certain distinctiveness and rustic beauty. Increased knowledge of these languages and an appreciation of the nomenclature derived from them are among the elements in the understanding of Canadian geography and history.

Indigenous Names for Aboriginal Places

On 1 January 1987, the municipality of Frobisher Bay became officially known as Iqaluit. The name, meaning 'place of fish' in Inuktitut, was long used by the Inuit for their community at this major centre of the eastern Arctic on Baffin Island.

Settlement in the area of Frobisher Bay has its roots in the establishment of a Hudson's Bay Company post on the south side of Frobisher Bay in December 1914, some 125 kilometres southeast of the present municipality. The post, actually a simple 3-metre by 4-metre hut, was moved up the bay to Fletcher Island in the spring of 1915. Five years later it was transferred across the 35-kilometre-wide Frobisher Bay to Waddell Bay. From 1921 to 1949, the post was located on Cormack Bay, some 70 kilometres southeast of present Iqaluit. In the fall of 1949, the Frobisher Bay post was moved to Apex Hill, 6 kilometres southeast of the United States air base, built near the head of the bay in 1942. In 1951, the area of the air base was officially named Frobisher, and that of the trading post was called Frobisher Bay. The latter became Apex in 1965.

In 1971, the municipal hamlet of Frobisher Bay in the area of the air base was incorporated. It became a village in 1974, a town in 1980.

In December 1984, the residents of Frobisher Bay voted 310 to 213 to have the name replaced by Iqaluit. This was subsequently confirmed by the Northwest Territories Executive Council, to take effect on the first day of 1987.

In 1965, the small Inuit community between the townsite and Apex Hill was officially named Ikaluit. As the community is within the town of Iqaluit, this variant spelling was dropped in 1989.

As to the name for the water feature, originally called Frobisher's Strait in 1576 by Martin Frobisher, there are no plans to change Frobisher Bay.

Recognition of Aboriginal community names in preference to non-Aboriginal names began in the western Arctic, where Tuktoyaktuk replaced Port Brabant in 1950. Vilhjalmur Stefansson reported that 'Tuktuyaktok,' meaning 'place where there are caribou,' was in use when he was there in the winter of 1907. However, the Hudson's Bay Company introduced the name Port Brabant in 1936, and this was adopted by the Geographic Board a year later. In 1948, postal officials requested the acceptance of

Tuktoyaktuk for the new post office, but HBC and the noted ethnologist Diamond Jenness recommended Tuktuk. The board disliked this popular variation of the Aboriginal name, and adopted Port Brabant for the post office. Two years later the board accepted the local name Tuktoyaktuk for the community and post office.

Much of the recent work in Aboriginal name restoration has been undertaken by the Commission de toponymie du Québec to reflect more closely the usage of the Inuit. Examples were the change of Port Harrison to Inoucdjouac (although Port Harrison remained the postal name), Notre-Dame-d'Ivugivic to Ivujuvik, and Notre-Dame-de-Koartac to Koartac. Even Port Burwell was changed to Killiniq by the Québec commission, although the community, since abandoned, was in the Northwest Territories (now Nunavut).

In 1979 and 1980, the commission decided that the actual names and the romanized spellings used by the Inuit in their own communities should take precedence. The following year Canada Post agreed to change its postal designations to conform to local preferences. The main changes that resulted in Québec were:

- Inukjuak, from Port Harrison (1908) and Inoucdjouac (1965)
- Kangiqsualujjuaq, from George River (1876) and Port-Nouveau-Québec (1965)
- Kangiqsujuaq, from Wakeham Bay (1930s) and Maricourt (Wakeham) (1965)
- Kangirsuk, from Payne Bay (1940s) and Bellin (Payne) (1965)
- Kuujjuaq, from Fort Chimo (1830–43, 1866) and Fort-Chimo (1965)
- Quaqtaq, from Koartak (1940s) and Notre-Dame-de-Koartac (1961) and Koartac (1965)

- Salluit, from Sugluk (1947) and Saglouc (1965).

As early as 1756 there was a trading post at Great Whale River on the east coast of Hudson Bay. The community was renamed Poste-de-la-Baleine in 1965, and the post office was changed to this in 1979. The Québec commission officially recognized the Inuit name Kuujjuaraapik in 1979 for the Inuit part of the village, and Whapmagoostui for the Cree part. Canada Post changed the post office to Kuujjuaraapik in 1992.

How well some of the Aboriginal names are being received throughout the Indigenous communities of the North remains to be seen. In arranging for air travel from Kuujjuaq to other communities, it has been said that some Inuit travellers will use the previous name (George River) to ensure that they get to Kangiqsualujjuaq, if that's where they're going, and not to Kangiqsujuaq.

Elsewhere in Canada, the trend of changing community names to those used by Aboriginal people, where they form the majority of residents, is slowly picking up.

In British Columbia, the Indian band council of Port Simpson requested in 1985 that its community be changed to Lax Kw'alaams. Meaning 'place of wild roses' in Tsimshian, this name had long been used by the Indigenous people. The change was officially made in July 1986, based on agreement by the names committee members for British Columbia and the federal Department of Indian and Northern Affairs. Canada Post also renamed its post office. Port Simpson itself was named in 1831 for Capt. Aemilius Simpson, who was then employed in the Nass River area in the marine service of the Hudson's Bay Company.

In 1989, the name of the hamlet of Eskimo Point, Nun., on the west coast of Hudson Bay, was changed to Arviat. In 1992, Spence Bay, Nun., at the south end of Boothia

Peninsula, became Taloyoak, and the Dene name Łutselk'e was substituted for Snowdrift, on the southeastern shore of Great Slave Lake. Coppermine became Kugluktuk and Fort Norman became Tulita, both in 1996.

By the end of the century there may be few English and French names of Aboriginal communities left in Canada. As reviews are made of each community name, don't be surprised to see Cambridge Bay become Ikaluktutiak, Coral Harbour change to Salliq, Hall Beach to Sanirajak, Resolute to Qausuittuq, and Chesterfield Inlet to Igluligaarjuk. Some other Dene community names may also be changed, such as Nahanni Butte to Tthenaagoo.

Since our adjustment to Tuktoyaktuk is complete, and since references to Inukjuak and Kuujjuaq can usually now be made without adding their former names, we shouldn't worry too much as more and more changes are made from English and French names to Indigenous ones.

Miramichi: Jacques Cartier in the 'Land of the Mi'kmaq'

One of the intriguing names of Eastern Canada is Miramichi. The name is best known as that of a river which drains a basin of 12,740 square kilometres in New Brunswick. Its principal branches are colloquially called the 'Sou'West,' the 'Nor'West,' and the 'Little Sou'West.' Miramichi is also a component of the longest geographical name in Canada: Lower North Branch Little Southwest Miramichi River.

In the form 'the Miramichi,' the name applies to a broad region of rugged terrain and dense woods from Boiestown to the city of Miramichi and beyond. The name occurs in the community Nelson–Miramichi, opposite Newcastle; and in the present community of Miramichi, east of Chatham, where a post office by that name existed from 1940 to 1970; a century earlier another post office by that name served the Chatham area.

Miramichi has even been applied three times beyond New Brunswick: to Pembroke, Ont., which briefly was called Miramichi in the early 1800s; to a brook and a small community south of Inverness on Cape Breton Island; and to a pond in Massachusetts.

One of the original 1867 House of Commons ridings was Northumberland–Miramichi. In 1988, it became simply Miramichi. Three of the fifty-eight provincial constituencies have it as part of their names: Miramichi Bay, Miramichi–Newcastle, and Southwest Miramichi.

Miramichi has its roots in the 1534 voyage of Jacques Cartier. Although his narrative makes no mention of it, maps derived from his historic voyages are the first to record it in the form of *Merchemay* (1541) and *Mecheomay* (1550). Similar forms appeared on later maps and documents, occurring as *Misamichy* on Samuel de Champlain's 1632 map, *Mizamichi* on Nicolas Sanson's 1656 map, and *Mizamichis* on Chrestien LeClercq's 1691 map.

Meanwhile, in 1632, Jean Guérard produced a map on which he inscribed *Grand Miramichi* (for today's Restigouche River) at the head of Chaleur Bay, and *Petit Miramichi* in the area of the present Miramichi. *Lustagooch* ('good river') and *Lustagoocheech* ('little good river') are the Mi'kmaq names for the Restigouche and the Miramichi, respectively.

The form with the 'r' on Guérard's map may have been a misprint, but it was repeated by Charles Lalement in 1645 and 1659, by Nicolas Denys in 1672, and by his son Richard, the first European settler in the Miramichi. Thereafter, map makers gave precedence to Miramichi.

In a thorough analysis of the roots of the name, William Francis Ganong concluded in 1926 that Cartier had learned it from the Montagnais in the area of Natashquan, opposite Île d'Anticosti on Québec's North Shore. On telling them where he had sailed, they probably said that Cartier had been in the 'land of the Mi'kmaq,' or, in their words, 'Maissimeu Assi.' This expression was then transformed into Mecheomay and related forms.

This derivation certainly seems to have more merit than several other origins given in various references, such as 'happy retreat,' 'ugly beaver,' 'place of many berries,' and 'river of many branches.'

Britain's redoubtable Lord Beaverbrook was raised in Newcastle, where his father, William Aitken, was a Presbyterian minister. When raised to the peerage in 1917, he intended first to call himself 'Lord Miramichi,' but the eminent Miramichi historian Louise Manny cautioned him that he might become known as 'Lord Merry Mickey,' and thus bring discredit upon himself and the region. He turned then to Beaver Brook, a small village north of Newcastle where he had once borrowed a fringed surrey from the Catholic priest for an Orange parade. Thus Max Aitken became Lord Beaverbrook.

The correct pronunciation of Miramichi often distinguishes the locals from the outsiders. With emphasis on the first and last syllables, it closely resembles 'MEER-a-me-SHEE.'

Saskatoon: A Home-Made Name from a Home-Grown Fruit

Saskatoon, Saskatchewan – a handsome prairie city founded on teetotalism – has a name that is pure Canadian.

Indeed, what could be more Canadian than Saskatoon? Its naming occurred during a religious service when a branch of delicious berries was presented to the leader of a new settlement on the banks of the South Saskatchewan River.

During the summer of 1881, the government of Canada advertised land on the Prairies for one dollar an acre. In Toronto, some abstemious individuals saw this as an opportunity to establish an alcohol-free community. They banded together to form the Temperance Colonization Society and applied for a block of half a million acres (200,000 hectares) straddling the South Saskatchewan. The following spring, the society's leader, Methodist missionary John Neilson Lake, and a number of society members boarded the Canadian Pacific train for Moosomin in what was then the District of Assiniboia. Prom there, they journeyed by wagon 430 kilometres to the northwest, arriving at their destination towards the end of July.

On 18 August 1882, Lake chose a site for the centre of the new colony near the present corner of Saskatoon's Broadway Avenue and Main Street, a short distance from the east bank of the South Saskatchewan. In his diary (written twenty years later), Lake noted that he planned to name the new settlement Minnetonka. He chose the name because Chief White Cap of a Siouan band at nearby Moose Woods recommended the site chosen, and because it was Siouan for 'big river,' a reference to the South Saskatchewan.

On Sunday, 20 August, Lake led a religious service at his camp. According to Adam Turner in his book, *Early Days in Saskatoon* (1932), Gerald Willoughby, a member of Lake's party, brought a branch of reddish purple berries to the service. It seems likely that the branch was given to Willoughby by one of the Cree chain bearers in a survey party that was laying out a plan for the town, and that the chain bearer told him about the importance of the berry in the Native diet and in making pemmican. Willoughby (who had become the party's interpreter of Cree and Siouan) informed Lake that the purple berries were called *misaskwatomina* in Cree.

To Lake, this sounded like *sas-ka-toon* and he announced this would be the name of the new settlement. The literal translation of the Native name is said to be 'fruit of the much wood.' Saskatoon berry is also called the early berry, serviceberry, and Juneberry *(Amelanchier alnifolia)*.

Towards the end of the twentieth century, a number of provincial historians and writers, including Bill Barry, author of *People Places: Saskatchewan and Its Names* (1997), cast doubt on the veracity of John Lake's story, observing that ripe saskatoon berries could not be found as late as 20 August. Barry was inclined to accept a story given by Edward Ahenakew in 1919 that the site, known as *manimkisâskwatân* in Cree, was where they cut saskatoon branches for arrow shafts. He concluded that it was 'likely that Lake had that phrase in mind when he chose the name.' Barry and others argued that it would be quite unlikely that Lake could have derived Saskatoon from *misāskwatōmina*. However, in 1810 geographer David Thompson spelled the name of the fruit *Sarskutum*, and in 1859 artist Paul Kane referred to *Sasketome* berries. Although fur trader Daniel Harmon rendered the Cree word as *Mi-sas-qui-to-min-uck* in his 1800 journal, and Sir John Franklin reproduced the word as *meesasscootoomeena* in 1820, most nineteenth-century references missed the first syllable. In May 2000 I heard a Cree scholar pronounce the word, and, although I was standing within two metres of him, I did not hear the first syllable, *mis*; but the second syllable, *sas*, was clearly heard.

In November 1999, I asked Clarence Peters, a horticulturalist in Regina with particular expertise with the saskatoon tree and its season of bearing edible fruit, whether ripe berries could be found in late August. He replied that it could have been a pear-shaped subspecies of the saskatoon that was found, which he said his family had collected just north of Saskatoon during his youth, and found them 'best from early to late August, when they were starting to wrinkle like a prune. Frequently we allowed them to ripen and dry on the trees and collected them during harvest, even towards the end of August. I think it is quite conceivable that the berries would have been picked in mid August.' As the story told by Lake was confirmed by other residents of Saskatoon 100 years ago, I see no reason for challenging it.

By the spring of 1883, the *Regina Leader* was carrying advertisements for the 'Temperance Colony, City of Saskatoon.' However, the settlement languished until 1890, when a railway from Regina to Prince Albert was rerouted to cross the South Saskatchewan at Saskatoon. Because it was easier to obtain water for the steam engines on the low flood plain along the west side of the river, the Saskatoon station was constructed there.

The first Saskatoon post office was built in 1884 on the east side of the river at the corner of Broadway Avenue and Eleventh Street. A new post office, called West Saskatoon, was established on the west bank in 1900 to serve the growing community there. Before long, the burgeoning west side had appropriated the name Saskatoon for its post office.

Thomas Copland, a leading merchant and land agent, was asked to resolve the confusion by devising a new name for the east-side post office. His proposal, as reported to the chief geographer of Canada, James White, on 3 November 1905, was to spell Saskatoon backwards, slightly modified, producing Nootaska. The Post Office Department amended it to Nutana, which Copland noted was a great improvement even though it obscured the origin of the name. Later, Adam Turner stated, and many others have repeated, that *Nutana* was a Siouan word for 'first born,' but such a claim came about through a misunderstanding by the deputy postmaster general in Winnipeg in 1901 when replying to Thomas O. Davis, M.P., who had submitted the Siouan word for 'first born.' In his reply, the postal official referred to Davis's submission as 'Nutana,' and others subsequently reached the same conclusion. The name likely submitted by Davis was Wenonah or Winona, celebrated by Longfellow in *The Song of Hiawatha*. The same epic poem provided the names Nokomis, Sask., and Wawanesa, Man. There are places called Winona in Ontario, Minnesota, and Indiana.

Nutana continued as a post office until 1919, when it became Saskatoon No. 1. Nutana remains today as the name of a neighbourhood between the river and the campus of the University of Saskatchewan. It also occurs in the names of twelve businesses and institutions, and in the name Saskatoon Nutana, a provincial electoral district.

In 1903, the villages of Saskatoon and Riversdale on the west side and Nutana were amalgamated to form the town of Saskatoon. Three years later, the now-expanding, self-proclaimed 'Miracle City of the West,' which even billed itself as the 'Fastest Growing City in the World,' was incorporated as a city in the newly created province of Saskatchewan. By 1908, three transcontinental railways – the Canadian Northern, the Canadian Pacific, and the Grand Trunk Pacific – converged on what became known as the 'Railway Hub of the West.' Prone to labels, Saskatoon also is called the 'City of Bridges,' having seven of them over the South Saskatchewan.

In 1921, Saskatoon had a population in excess of 25,000 people, and by 1961, the figure stood at 95,526. The 1986 census recorded Saskatoon as the largest city in Saskatchewan, when its residents numbered 177,641 to Regina's 175,064. By the end of the century its population surpassed 220,000.

Saskatoon is an authentic Canadian name acquired from the immediate surroundings of the city. Its uniqueness is emphasized by the tale of two British visitors travelling west by rail. Wondering where they were when the train stopped after crossing a wide river, one of the travellers got off the train to inquire and was told: 'Saskatoon, Saskatchewan.' Reboarding the train, the traveller announced with disappointment: 'They don't speak English here.'

The name Saskatoon also occurs elsewhere in Western Canada. There is a Saskatoon Lake in both northeastern Saskatchewan and northwestern Alberta. In the same area of Alberta are Saskatoon Hill, Saskatoon Lake, and Saskatoon Island Provincial Park. Across the border in British Columbia is Saskatoon Creek. In southern Alberta, near the Crowsnest Pass, is Saskatoon Mountain.

Both the place and the name that John Lake created on the western landscape were symbolically brought to Toronto in the 1950s by the naming of Saskatoon Drive, now in the central part of Toronto's Etobicoke. One can imagine this would have pleased Lake, who died in 1925 in Toronto at the age of 91.

Medicine Hat: Mystery, Romance, and Hints of Magic Below

The story of how Medicine Hat got its name is so intertwined with folk tales that the truth can no longer be authenticated. The most publicized story is that a Cree medicine man had his hat blown into the South Saskatchewan River during a battle with the Blackfoot. When that happened, the Cree warriors fled, leaving the Blackfoot to declare the site 'Saamis,' or 'place of the medicine man's hat.'

The Geographic Board report of 1919 published that account, and it was repeated in 1928 in Robert Douglas's *Place Names of Alberta*. However, both these publications, so as not to show any preference among the tales, presented four other stories:

- A hill east of Medicine Hat was said by Walter Johnson, the first settler there in 1882, to be shaped like a medicine man's head-dress.
- A medicine man took a fancy to a hat taken from a white settler who, together with his companions, had been massacred.
- A brave was rewarded with a fancy head-dress by a medicine man after he rescued a girl.
- A chief had a vision of a figure rising out of the river and wearing the plumed hat of a medicine man.

When this essay was first published in *Canadian Geographic* in 1984, I reported that there were at least two other stories that explained the origin of the name: (1) A Blackfoot warrior was directed in a dream by a figure wearing a head-dress to drown his betrothed so that he might become a great chief. He drowned her, became a powerful chief, and declared the site 'Saamis'; (2) William Van Horne, the renowned builder of the Canadian Pacific Railway, was told in 1883 by two Métis hunters that a Cree brave named Thunder Bear ran off in 1862 with the young wife of an old Cree chief. The wife is said to have dreamed that if Thunder Bear would give the old chief a war bonnet with the tail feathers of seven eagles, he would leave Thunder Bear in peace at the site where the river (South Saskatchewan) comes closest to the Cypress Hills. When Van Horne was told the story he is said to have remarked: 'That hat was good medicine to appease the irate chief,' and thereupon named the place Medicine Hat.

In 1994 Marcel M.C. Dirk published a small booklet entitled *But Names Will Never Hurt Me,* and reported that there were at least twelve explanations for the origin of the name, with at least six of them displaying only subtle variations of a single theme.

Shortly after Medicine Hat achieved city status in 1906, Rudyard Kipling paused there on a transcontinental trip. On learning that the area was sitting on top of a huge natural-gas field, he declared that it had 'all Hell for a basement with Medicine Hat for a trap door.'

In 1910, some newcomers to the community suggested that the city should have a more prosaic name, such as Gasburg or Smithville. Members of the city's Cypress Club were outraged, and delegated Frank Fatt, the postmaster, to appeal to Kipling for support in retaining the original name. Kipling responded forthwith from England:

'To my mind,' he wrote, 'the name of Medicine Hat ... echoes the old Cree and Blackfoot tradition of red mystery and romance that once filled the prairies. Also it hints at the magic that underlies the city in the shape of your natural gas.

'Believe me, the very name is an asset, and as years go on will become more and more of an asset. It has no duplicate in the world; it makes men ask questions, and ... draws the feet of the young men towards it. It has the qualities of uniqueness, individuality, assertion and power. Above all, it is the lawful, original, sweat-and-dust-won name of the city, and to change it would be to risk the luck of the city, to disgust and dishearten old-timers, not in the city alone, but the world over, and to advertise abroad the city's lack of faith in itself.

'Men do not think much of a family which has risen in the world, changing its name for social reasons. They think still less of a man who because he is successful repudiates the wife who stood by him in his early struggles. I do not know what I should say, but I have the clearest notion of what I should think of a town that went back on itself ...

'Forgive me if I write strongly, but this is a matter on which I feel keenly ... I have a huge stake of interest and very true affection in and for the city and its folk.

'... The *Calgary Herald* writes [that a new name would have] a sound like the name of a man's best girl and looks like business at the head of a financial report.

'But a man's city is a trifle more than a man's best girl. She is the living background of his life and love and toil and hope and sorrow and joy. Her success is his success; her shame is his shame; her honour is his honour; and her good name is his good name.

'What, then, should a city be rechristened that has sold its name? Judasville.'

Canada Has Its Mississippi River Too

Picture it. I am driving over a long bridge on Ontario's Highway 7, some 60 kilometres southwest of Ottawa, when my out-of-province passenger stops talking in mid-sentence, and asks incredulously: 'Did that sign say Mississippi River? Is this near the head of North America's greatest river?' The answers are yes and no.

Canada's Mississippi River, a tributary of the Ottawa,

rises in the Madawaska Highlands of Central Ontario, and, in its 201-kilometre length, drops 323 metres. In its upper section, it flows south through the picturesque Upper and Lower Mazinaw lakes, in Bon Echo Provincial Park. After draining the smaller Marble and Georgia lakes, it swings east to flow through Kashwakamak, Crotch, Dalhousie, and Mississippi lakes. Just east of the town of Carleton Place, it turns northwest, and after tumbling over falls and rapids at Appleton, Almonte, Blakeney, Pakenham, and Galetta, its main course flows into Lac des Chats, a reservoir on the Ottawa River. Its second outlet, the Mississippi Snye, flows east for 6 kilometres, emptying into the Ottawa below the Chats Falls dam and generating station.

Although the double outlet of the Mississippi was well known to explorers and settlers from the early 1800s, Canadian topographical maps and sailing charts until the 1970s incorrectly identified the final 2 kilometres of the main outlet plus the snye as the Ottawa River (South Branch)!

The name Mississippi is puzzling for this particular river. While the name suggests in some Algonquian languages, such as Cree, the biggest river among surrounding rivers, it is not only smaller than the adjoining Madawaska River, it is mere mite beside the mighty Ottawa (1,190 kilometres), which itself is known in the Algonquin language as Kitchisippi, 'great river'; and is a barely a trickle compared with the magnificent Father of Waters, at 3,780 kilometres, which drains the American plains into the Gulf of Mexico. It might be concluded that the name of Canada's Mississippi is really an error for another name in Ojibwa, Mississauga, or Algonquin, with the early explorers and map makers having got it wrong, much as they evolved Madawaska from Matouachita.

Samuel de Champlain ascended the Ottawa River in 1613, but made no mention of a river entering the Ottawa above Chats Falls, and did not portray it on his subsequent maps. Rivers flowing northeast into the Ottawa were not shown or named on maps until the late 1600s. On Guillaume Delisle's 1703 *Carte du Canada*, four were identified: their present-day river names are South Nation, Rideau, Madawaska, and Bonnechere. But during the whole of the eighteenth century nary a squiggle was made on maps to identify the Mississippi, nor was there apparently any mention of it in reports and published materials.

When Britain took over the area of present-day Eastern Ontario after 1763, its map makers portrayed only well-known features, with minor features ignored if they had not been surveyed. Until surveyors laid out the townships northwest of the Rideau River in the second and later decades of the nineteenth century, maps of that area were intentionally left totally blank.

In May 1816, Alexander McDonell, a superintendent of locating settlers in Upper Canada, reported to the secretary of the lieutenant-governor that he had 'been informed by Indians and others that in the rear of the River Tay there was a much larger River which empties into the Ottawa.' He directed surveyor William Graves to investigate the river. Graves reported that it was fine, and the land between the two rivers was of an excellent quality. Neither of them identified it by name.

The earliest reference to the river is in Graves's 1816 diary of his survey of Bathurst and Drummond townships. On 16 July, he referred to it as the *Mifsippy River.* Elsewhere in his diary and in his field book he used four other spellings. In a field book by one of his assistants, the name is written *Missisippi River.* The first mention of *Mifsifspee Lake* (upriver from present-day Carleton Place) is in Willson Conger's 1817 survey diary of Beckwith Township. Also that year, John Ryder mentioned *Maſsassippa River* in his survey diary of Beckwith and Goulbourn townships. In 1818, a government map of the eastern part of Upper Canada identified *R^r Miſsiſsippy*, but did not name the lake, although its outline was accurately portrayed. The modern spelling for the the river was first published on the 1818 *Sketch of the Rideau Settlement*, but the compiler assigned *Miſsissippi Lake* to present Bennett Lake, west of Perth. An 1821 map of Bathurst District not only spelled *Mississippi River* and *Mississippi Lake* as we write them now, but also gave the name *Mississippi Lake* to Bennett Lake, and, as well, called Silver Lake on present-day Highway 7 *West Mississippi Lake.*

Speculation that the name Mississippi may have evolved from Mazinaw was raised in 1968 by Carleton Place historian Howard Brown. Derived from one of the Algonquian languages, Mazinaw refers to the painted pictographs on Mazinaw Rock on the east side of Upper Mazinaw Lake, according to pictographer Selwyn Dewdney (1909–1979), the father of Canadian scientific rock study. It might not be too difficult, on hearing *Mazinaw-zeebi*, 'painted image river,' to render it by the more familiar Mississippi River. Perhaps its name could have been derived from the name of the Mississauga, a division of the Ojibwa, who migrated from the north shore of Lake Huron to the area of the upper Mississippi River in the mid 1700s. In 1827, Hudson's Bay Company factor Charles Thomas, based at Chats Falls, referred to the river as *Tomississippi* and *Tomissipii*, but neither appears to have an obvious meaning in Ojibwa or Algonquin, according to the nineteenth-century dictionaries of those languages.

In 1823, engineer Samuel Clowes and surveyor Reuben Sherwood examined the course of the Mississippi to determine its suitability as a route for the projected canal from Lake Ontario to the Ottawa River. Although finding it a fine and copious stream, it was concluded in the report to the lieutenant-governor that 'the bed of the Miſsiſsippy was far too elevated, and that as the lockage to attain and descend from the Summit Pound would be enormously expensive no Canal would be practicable in that direction.'

The Kingston and Pembroke Railway was built from Kingston to Renfrew between 1872 and 1884, reaching the Mississippi River in 1878. When a post office was named Mississippi Station on the south side of the river in 1879, the community of Mississippi, with a store, hotel, and meeting hall, grew up around it. In 1959, the station was closed, but the post office continued in use until 1987, and Mississippi remains a small, but viable community.

In 1884, the Mississippi River figured in an important piece of British law. The most prominent lumber baron on the river in the 1870s and '80s was Peter McLaren of Perth. He had a lumber mill in Carleton Place that cut some 100,000 feet of lumber a day, and had constructed a number of dams, sluices, and slides along the length of the Mississippi. By 1886, McLaren employed some 450 men in 10 shanties, as well as a number of other part-time wood cutters. He believed the improvements he had made on the Mississippi entitled him to prevent anyone else running logs down his slides. Boyd Caldwell, another lumber baron with limits on the Mississippi and its tributary, the Clyde, won the approval of the Ontario government three times to use the Mississippi waterway for his logs and square timber, but Sir John A. Macdonald's Canadian government rejected the provincial permits. Britain's Privy Council supported Caldwell's claim by passing the Rivers and Streams Bill in 1884. Since that time, it has been unlawful for anyone in Canada to have a monopoly over a navigible waterway.

The 20-kilometre-long Mississippi Lake is a very popular destination for vacationers, with some 1,300 summer cottages around its shores. The first steamboat on the lake in 1867 was aptly named *The Mississippi*. On 1 January 1998 the town of Almonte and the townships of Pakenham and Ramsay were amalgamated to create the town of Mississippi Mills.

The source of the Mississippi is only a few metres east of the source of the Little Mississippi River, which drains the intricately shaped Weslemkoon Lake to the north for 25 kilometres, where it joins the York River, a major tributary of the Madawaska. In the days before European contact, the two river systems likely formed important components of canoe travel between Lake Huron and the upper St. Lawrence River. There is a strong tradition that a great battle was fought in the area of the headwaters of the Mississippi. And as well, it is believed the same area may be where Champlain spent the winter of 1616 with the Huron.

Throughout its length, the Mississippi River possesses charm and four-season splendour for the tourist, the cottager, the back-to-nature lover, and the basin's immense variety of farmers, woodsmen, merchants, professionals, and townspeople. It seems surprising, then, to find the first evidence of the name as late as William Graves's diary of 1816, when the neighbouring Rideau, Madawaska, and Bonnechere rivers had been recorded for more than 100 years.

In Canso, Survival Is a Tradition

Canso, N.S., knows how to survive adversity. The burdens it has shouldered in its long history – isolation, harsh climate, shallow soils, unreliable fish stocks, depressed fish prices, oil spills, drownings, and fish-plant closings – have strengthened its fibre and durability. As recently as 1990, the community had to fight for its survival through a yearlong campaign to reopen its fish plant: the tenacious effort restored 500 jobs and impressed Canadians from coast to coast.

Canso is located at the seaward end of Chedabucto Bay's south shore, 200 kilometres northeast of Halifax. Several islands shelter its harbour from the open waters of the bay and the Atlantic Ocean. The town has a population of 1,230, with another 500 in the adjacent communities of Hazel Hill, Tittle Road, and Durells Island.

European fishermen were attracted to Canso's sheltered harbour from the first decade of the 1500s. For the next 200 years, French, Portuguese, and Basque fishermen built wharves, fish flakes, and shelters on the shores of the islands and mainland, and fished for cod. The English took over Canso when most of Acadia (now mainland Nova Scotia) was transferred to Britain in 1713 by the Treaty of Utrecht. Fishermen, merchants, and traders from New England – with the protection of a British garrison – developed a prosperous fishery on Grassy Island in Canso Harbour, and English records declared it to be 'the best and most convenient fishery of any part of the King's dominions.'

Those advantages also marked Canso as a military target. In 1744, French forces from Louisbourg, Cape Breton Island, captured and burned the settlement, calling down the wrath of both the English and the New Englanders. A British fleet with New England militia attacked and captured the Louisbourg fortress the following year – after training on Grassy Island. The island, which has remained largely undisturbed for nearly 250 years, is now a national historic site.

Canso was razed again during the American Revolution when privateers, led by the notorious John Paul Jones, attacked the community in 1776. The devastation discouraged settlement, and even twenty-five years later there were only about six families living in the area. Permanent settlement finally began around 1812, a 'way office' for mail transfer was established in 1834, and by mid-century the population had increased to 250 families. Two cable companies operated their headquarters in Canso and neighbouring Hazel Hill – for communications with Europe – from 1881 to 1961, providing a source of employment aside from fishing and fish packing. Canso was incorporated as a town in 1901.

Canso's name is derived from the Mi'kmaq language. Rev. Silas Rand, a nineteenth-century authority on the language, gave the Mi'kmaq word as *camsook* or *kamsook*, meaning 'opposite the lofty cliff.' In a 1914 study of Canso, Dr. William F. Ganong, a renowned authority on Maritime Native names and their meanings, said the root *kam* meant 'across' or 'beyond,' and *sook* meant 'cliff.' Ganong concluded that the line of cliffs on the south shore of Chedabucto Bay provided the source for the name, and this conclusion has been repeated in a number of studies of place names. However, residents suggest the name may owe its origin to a mirage, seen from the Canso shore, of an impressive height of land on Isle Madame, 13 kilometres to the north. Although this hill is only 65 metres high at Cape Auguet, its image – seen over water – seems magnified under certain atmospheric conditions that distort light rays.

The Mi'kmaq origin was noted as early as 1609 by the French writer Marc Lescarbot in his *Histoire de la Nouvelle France*, but that did not prevent later searches for a European origin. At least one writer linked the name to the Spanish *ganso* for 'goose.' Others have suggested it was derived from the name of a French explorer called Canse, plus *eau* for 'water' – a mystifying hypothesis, since there is no evidence such an explorer ever lived. Lescarbot was the first to use the French transliteration – 'Campseau' – to describe the harbour, the present Chedabucto Bay, and the narrow strait separating Cape Breton Island from the mainland. Samuel de Champlain referred to 'Port de Canseau' in his *Voyages* (1613) in reference to the harbour. In the 1632 edition of his *Voyages*, 'Canseau' is used for the harbour and strait, 'Campseau' for the bay. Colonizer Nicolas Denys used 'Campseaux' for all three features in his 1672 account.

The use of 'Canceau' for present Cape Canso (the outer point of Andrew Island, 5 kilometres southwest of the town) first appeared on hydrographer Jean-Baptiste-Louis Franquelin's 1686 map of Acadia. Later editions of the map and one by Guillaume Delisle (1703) spelled the cape's name 'Camseau.' Map maker Jacques-Nicolas Bellin used 'Canseau' on his 1744 map of Acadia for the harbour, cape, and strait, and this spelling prevailed thereafter on French maps.

The earliest appearance of the name in an English document is on a Herman Moll map of 1715, when he used 'Canseaux' for the strait. The change to today's spelling came in 1749, when Nova Scotia surveyor general Charles Morris substituted 'Canso' for the French version, and this prevailed as the official form.

Strait of Canso, for the body of water between Cape Breton Island and the mainland, was first used in 1776 by chart maker Joseph F.W. DesBarres, and it ultimately replaced 'Gut of Canso' on maps and marine charts by about 1865.

Canso has much to offer inquisitive travellers, even those who arrive in search of the Canso Causeway, which is some 100 kilometres northwest of the town, by road. The causeway joining the mainland to Cape Breton

Island was built in 1955, and perhaps should have been called the Strait of Canso Causeway, or Road to the Isles Causeway, to avoid confusion with the distant town.

Canso's boosters are proud to show visitors the Whitman House Museum, the Seafreez fish-packing plant, the graceful Star of the Sea Roman Catholic Church, the Grassy Island visitor interpretation centre, and the impressive seamen's memorial. Knowledgeable residents talk proudly about the first mass celebrated in Nova Scotia (1611) and the first Anglican church built in Canada (1720) – both in Canso. They recall the survival instincts of the little town where, in the Dirty Thirties, Fr. Jimmy Tompkins publicized the poverty of Maritime fishermen: he later introduced the concept of credit unions to the province. They also have no doubt that Canso will continue to be a proud, 'can so' community.

Nanaimo: A City and a Sweet Dessert

No Canadian place name is more celebrated in Canadian bakeries, kitchens, and social gatherings than the Vancouver Island city of Nanaimo. Whether it be at a monthly meeting of a Women's Institute in the Maritimes an evening card party in Southern Ontario, in your favourite cookbook, or a packaged square mix in the local grocery store, the Nanaimo Bar is ubiquitous.

The name Nanaimo owes it origin to a Coast Salish confederacy of five bands, the Koltsiowotl, the Ksalokul, the Yesheken, the Anuenes, and the Tewetken, who wintered at their village of Stli'ílep, in the area of Departure Bay on the north side of the present city. The Nanaimos described their joint community as *Snenéymexw*, meaning 'the whole,' 'big strong tribe,' or 'the great and mighty people.' Their territory extended from Boat Harbour,

17 kilometres southeast of the centre of the city, to Horswell Bluff, 6 kilometres to the north, and extended inland up the Nanaimo River, called by the Nanaimos Quamquamqua, 'swift, swift water.'

The first documentary evidence of the name Nanaimo occurs in an 24 August 1852 letter from Vancouver Island governor and Hudson's Bay Company factor James Douglas to company clerk Joseph McKay 'to proceed, with all possible diligence to Wintuhuysen Inlet, commonly called Nanymo Bay, and formally take possession of the coal beds lately discovered.' Douglas used the current spelling in an 13 October 1852 letter to McKay, and the following spring Commander James Prevost referred to the 'Harbour of Nanaimo' in a despatch sent to the British Admiralty.

In 1791, Spanish sailing master José Maria Narvaez, a member of Francisco Eliza's expedition, named the 15-kilometre-long stretch of bays, coves, and passages from Northumberland Channel on the southeast to Departure Bay on the northwest *Bocas de Winthuysen*, for Spanish rear admiral Francisco Xavier de Winthuysen. After 1853, all references to the Spanish admiral with Flemish roots disappeared from the toponymy of the West Coast.

In 1851, a Nanaimo called Chewechkan learned that a blacksmith at Fort Victoria was using coal imported from England, and informed Joseph McKay that similar coal was available near the shore of his community. When Chewechkan returned in the spring of 1852 with good-quality coal, he was rewarded with the title of Coal Tyee, or 'coal chief,' which he proudly used until his death in 1884. He is recalled in the street name Coal Tyee Trail.

Governor Douglas visited the site in 1852, and, with the help of the Nanaimos, mined 50 tons of coal. He reported that 'the discovery has afforded me more satisfaction than I can express.' Joseph McKay purchased almost 6,200 acres from the Nanaimos for the price of 668 Hudson's Bay blankets. Douglas and McKay found the Nanaimos, who are reported to have had 943 men then, quite peaceful and keen to be involved in the mining of coal, but they deemed it necessary to build a bastion to protect the site from the warlike Haida raiders from up the coast.

The community around the mine site was named in 1854 Colviletown, for Andrew Colvile, governor of the Hudson's Bay Company. The name was often misspelled as Colvilletown and Colville; Town letters were addressed 'Colville Town, Nanaimo, V.I.' until 1860, after which Colviletown disappeared.

The first group of twenty-one miners from the north end of Vancouver Island and the California gold fields arrived in Nanaimo in 1852. On 27 November 1854, after sailing 140 days from London via Cape Horn, a group of twenty-one Staffordshire miners, with their wives and forty-two children, came ashore. In 1862, the Hudson's Bay Company divested itself of its coal-mining interests by selling out to the Vancouver Coal Mining and Land Company, based in London. (It would be another twenty-three years before Vancouver would be identified with a mainland city.) That year, six tons of Nanaimo coal was exhibited at the London World's Fair. In 1950, mining operations ceased. During the 100-year-period, total coal production had surpassed 50 million tons.

In 1863, to facilitate the mining operations, the first rail line in Western Canada was built at Nanaimo. The name of the city is reflected in the Esquimalt and Nanaimo Railway, built by Robert Dunsmuir, with Sir John A. Macdonald driving the last spike on 13 August 1886. Dunsmuir's son James sold the railway to the Canadian Pacific in 1913 for $3,000,000.

Nanaimo was the name of an 800-ton barque built in

1882 by Nanaimo Sawmills, but it was soon sold to Hong Kong interests. The ferry *City of Nanaimo* served the Vancouver-to-Nanaimo run from 1891 to 1912. Canadian Pacific operated the *Princess of Nanaimo* on the same run for two years, 1951 and 1952. She was followed by BC Ferries' *Queen of Nanaimo*, which is still in service between Tsawwassen, on the mainland, and the Gulf Islands. In 1940, the corvette H.M.C.S. *Nanaimo* was launched from Yarrow Shipyards in Victoria and served on Atlantic convoy duty throughout the Second World War.

A second community developed across the harbour on Newcastle Island, named in 1853 because of the mining of coal there by the Hudson's Bay Company. In 1874, Nanaimo and the Newcastle townsite were joined to form the city of Nanaimo, the third city incorporated on the West Coast. It then had 1500 people. Between 1927 and 1954, the city was enlarged some ten times by annexing the neighbouring communities of Northfield, Departure Bay, Wellington, and Chase River. At the beginning of the twenty-first century, the 'hub city,' as the mid-island city is called, has a population of some 75,000.

In the 1920s, there were forty Japanese Canadian herring salteries in the Nanaimo area. Perhaps it was the significant presence of Japanese fishermen before the Second World War that has led more than one person to speculate that Nanaimo was derived from the Japanese words for 'seven potatoes.'

In 1985, Nanaimo's mayor Graeme Roberts advertised a search for the 'Ultimate Nanaimo Bar.' Among the 100 different recipes submitted, local resident Joyce Hardcastle's was the unanimous favourite among the judges, and it is her recipe that city hall and the chamber of commerce hand out. Among her secrets were the use of unsalted cultured butter and the substitution of almonds for walnuts, as recommended in earlier recipes.

In a chamber of commerce report on the history of the bar, Nanaimo resident Phyliss Milligan is quoted as saying a recipe for Chocolate Fridge Cake was copied about 1936 from a *Vancouver Sun* newspaper, and passed among a group of ladies. The group formed the Harewood Ladies Auxiliary in 1946. When they compiled a cookbook the delicious Chocolate Fridge Cake was called the Nanaimo Bar. Well-known cookbook writer Anne Lindsay spoke to Mrs. Mulligan in 1984 about the recipe, but was told both the clipping and the cookbook had vanished. Searches by *Vancouver Sun* food editors failed to provide any record of the Chocolate Fridge Cake. Ms. Lindsay was quoted in a 1984 article by Carol Ferguson in *Canadian Living* that the many variations of the recipe in North America have their source in a European pastry.

There are many contradictory stories about the true origins of the pastry that ultimately earned the title Nanaimo Bar, with New Yorker, New York Dream Bar, New York Slice, London Fog Bar, Edmonton Esk, and Domino being among either its antecedents or its parallels. Suffice it to say that it is a very popular dessert in the city of Nanaimo. The city's official mascot is Nanaimo Barney, shaped like a giant walking square.

There are two outlets on Nanaimo's Commercial Street that specialize in the production of the bar. The Scotch Bakery advertises itself as the 'Home of the Famous Nanaimo Bar.' The Nanaimo Bar and Cookie Company produces several varieties of the bar, including mocha, cappuccino, and one with an extra-thick chocolate top layer.

Many recipe books included a version of the Nanaimo Bar in the 1950s. By 1960, Bird's Custard Powder, one of the essential ingredients, had a recipe for the bar on its cans. Western sugar companies printed recipes on their bags after 1965.

Joyce Hardcastle's Ultimate Nanaimo Bar

Bottom layer:
125 ml (1/2 cup) unsalted butter (European-style cultured)
50 ml (1/4 cup) granulated sugar
75 ml (5 tbsp.) cocoa
425 ml (1 3/4 cups) graham wafer crumbs
1 egg, beaten
125 ml (1/2 cup) finely chopped almonds
250 ml (1 cup) unsweetened coconut

Melt the first three ingredients in the top of a double boiler. Add egg and stir to cook and thicken, about one minute. Remove from heat, and stir in crumbs, coconut, and nuts. Press firmly into an ungreased 20 cm^2 (8-inch-square) pan.

Middle layer:
125 ml (1/2 cup) unsalted butter
40 ml (2 tbsp. plus 2 tsp.) cream
30 ml (2 tbsp.) vanilla custard powder
500 ml (2 cups) icing sugar

Cream together butter, cream, custard powder, and icing sugar. Beat until light. Spread over bottom layer.

Top layer:
4 28–gram (1–ounce) squares semi-sweet chocolate
30 ml (2 tbsp) unsalted butter

Melt chocolate and butter over low heat. Cool. When cool, but still liquid, pour over second layer, and chill in the refrigerator. Lightly mark into 2.5-cm (1-inch) squares before chocolate hardens completely.

We Have 'Honoured' the Moose 662 Times!

Naming geographical features for wild animals was one of the commonest ways of differentiating elements of the landscape in the early exploration and settlement of North America. One animal that attracted much attention was the moose, a large, somewhat grotesque member of the deer family, similar to the European elk. The Algonquian-speaking tribes called it *moos*, meaning 'browser,' describing the moose's habit of stripping the tender bark and low branches of trees.

When the English explorers encountered the animals in the early 1600s in New England, they initially described them as 'moose-deer.' Eventually the shortened form 'moose' was used throughout the continent for the rather graceless animal with ungainly legs and drooping muzzle. The same animal acquired the identity of *orignal* in French, a Basque word for what the French call *cerf* ('red deer') in both France and Canada.

As well as achieving prominence on coats of arms, coinage, and bottled beverages, the moose has been 'honoured' in the names of 662 official names in Canada, concentrated in a wide band across Northern Ontario (166) and central Québec (109), and in mainland Nova Scoria (117).

The oldest 'moose' name in Canada may be Moose River in Northern Ontario, known in Cree as Mons Sipi. The area was once occupied by an Algonquian-speaking tribe called Monsoni ('moose people'). Thomas Gorst mentioned 'Moose Sebee' in his 1670–1 journal. On Moose Factory Island, 20 kilometres from the river's mouth, the Hudson's Bay Company established in 1673 its second trading post, Moose Fort (the first was Fort Rupert). In 1730, the post was moved farther upriver on the island, and today is the home of approximately 1,000 Cree.

Opposite Moose Factory Island on the west bank of Moose River, where Revillon Frères once had a trading post called Moose River Post, is Moosonee, a town of 1,250 people.

Moosonee, from the Cree name for Moose Factory, became the terminus of the Temiskaming and Northern Ontario Railway (now the Ontario Northland) in 1932. It is the destination of the popular summer tourist excursion, the Polar Bear Express. The Diocese of Moosonee was established in 1872 by the Anglican Church at Moose Factory. The diocesan centre is now based in Schumacher, a community in the city of Timmins.

Moose Jaw, Sask., has often been called the most extraordinarily crazy name for a town in the world. Moose Jaw River, a tributary of the Qu'Appelle River, was first recorded as Moose Jaw Creek by John Palliser in 1857.

Suggestions that the name commemorates the repair of a cartwheel with a moose's jaw by a titled Englishman (who may have been Lord Dunsmore or the Earl of Musgrave or Lord Milton) are considered to be mythical. The shape traced by the river as it flows, first, northwest, then, within the city of Moose Jaw uncannily resembles the outline of a moose skull. One of the earliest settlers to arrive in the area of the city in 1882, James (later Senator) Ross claimed that the formation of the hills as seen from the river's mouth had the shape of the jaw of a moose. In his *People Places: The Dictionary of Saskatchewan Place Names* (1998), Bill Barry has concluded that 'the name ... almost certainly comes from the Cree phrase *moscâstani-sipiy,* referring to an area along the river where the winter winds are less harsh.'

Moose Jaw was incorporated as a town in 1884, and as a city in 1903. The Geographic Board approved the forms Moosejaw Creek and Moosejaw for the town in 1901. In

1931, the board corrected its record of the city's statutory name to Moose Jaw. In 1967, the name Moose Jaw River replaced the earlier decisions of Moosejaw Creek (1901) and Moose Jaw Creek (1952).

The town of Moosomin, 220 kilometres southeast of Regina on the Trans-Canada Highway, received its name as a CPR station in 1882. It was given in honour of a prominent Cree leader, whose name is also commemorated in the name of Moosomin Indian Reserve north of North Battleford, Sask. The name is also the Cree word for mooseberry, a low-bush berry of the Prairie provinces.

In the early 1800s, the site of Moosomin was the crossing of a trail leading to Moose Mountain, a prominent elevation in southeastern Saskatchewan. In 1947, a federal electoral district was named Moose Mountain. In 1974, it was joined to Qu'Appelle to become Qu'Appelle–Moose Mountain, and, in 1988, it was renamed Souris–Moose Mountain.

L'Orignal, an attractive community in the Township of Champlain, has a population of 2,164. It is the county seat of the United Counties of Prescott and Russell and is located on the Ottawa River almost midway between Ottawa and Montréal.

The name may be traced to the Seigniory of Pointe-à-l'Orignal, granted to François Prévost in 1674. The point, east of the village, possibly named by voyageurs, is now called Grants Point. It is variously said to be the site where a moose or deer skeleton was discovered, or where the moose crossed the Ottawa River at this narrow place above the storied Long Sault, made famous as the last stand of Dollard des Ormeaux.

In many of the early land records, the name was also spelled L'Orignac. Frequently the name of the village has also been misspelled L'Original, even on postmarks.

Moose Creek is a small community 46 kilometres

southwest of L'Orignal. The small creek, a tributary of the South Nation River, is said to have a place that rarely freezes where moose used to gather in the winter. The current post office sign fuses the two words into Moosecreek.

Moose River Gold Mines, about 65 kilometres northeast of Dartmouth, N.S., were mined from the 1870s to the 1920s. The place became famous in 1936, when three Toronto men, exploring the possibility of reopening the mines, became trapped by a cave-in. After eleven days, while the ordeal was followed by millions through newspaper accounts and the first continent-wide use of radio broadcasts, two of the men were brought out alive. Nearby is Mooseland, where a post office was opened in 1877. Its current population is about 80.

Nova Scotia had two post offices called Moose River, one in Pictou County from 1874 to 1943, the other in Cumberland County from 1882 to 1950. The population of each is now smaller than 25. Near Digby is a small stream called Moose River, identified as 'R de l'Orignac' on Lescarbot's 1609 map.

Some moose were transferred from the Maritimes to the Island of Newfoundland in the 1870s and at the turn of this century. Their occurrence likely accounts for the seventeen geographical features with moose, mostly between Corner Brook and Grand Falls–Windsor.

The moose – an imposing beast of the Canadian woods and wilderness – is prominently reflected in the names of geographical features, from Moose Ear Pond in Newfoundland to Moose Bath Pond in B.C. and Moosehide Hills in the Yukon. The names – 662 of them – reveal the distinctive character of Canada's names.

Examining the Names of Particular Places

Cupids, Nfld.: Britain's First Canadian Colony

During a May 1991 visit to Newfoundland, I had the pleasure of touring the old town of Cupids. It has the distinction of being the first year-round British colony in what is now Canadian territory, having been founded by John Guy as the Seaforest Plantation in August 1610.

Guy's achievement is little known outside Newfoundland, and even within the province other events – the 1,000-year-old Viking settlement at L'Anse aux Meadows; Marconi's 1901 transatlantic radio broadcast at Signal Hill in St. John's – have received greater attention.

Guy was a city councillor in Bristol and a master of the Bristol Society of Merchant Adventurers. In 1608, the same year Samuel de Champlain founded Québec City in the name of France, Guy made a reconaissance of Conception Bay. On his return to England, he persuaded forty-eight investors to join him in a company to establish a plantation on the west side of the bay. In May 1610, the Privy Council endorsed his proposal, with the condition that his colony not interfere with the established fishery.

In early August, a ship carrying Guy and thirty-nine artisans – carpenters, stonemasons, shipwrights – sailed into the 'harbour here called Cuperres coue,' on a short arm of Bay de Grave, 35 kilometres west of St. John's. According to local history, the site was proposed by a man named Daw, whose ancestors reportedly had a fishing operation as early as 1522 at Ship Cove, on the north side of Bay de Grave.

Cupers Cove, as Guy later wrote it, likely owes its name to the practice of fish packers sending coopers to the well-treed slopes to cut wood for staves to make barrels for pickling fish. Cuperts coue, Cubitts Cove, and Coopers Cove are early variations of Cupids Cove, which came into use on maps and charts in the last quarter of the seventeenth century, after having been used in 1630 by the first governor of Nova Scotia, Sir William Alexander.

Selecting a site beside a small brook near the head of the cove, Guy set his artisans to work building a house, workshop, storehouse, forge, and barn, all enclosed by a stockade measuring 27 metres by 37 metres and protected by three cannons. They also built a wharf and a store on the lagoon at the head of the cove, as well as six fishing boats and a 12-ton shallop.

In October, Guy sent the company a glowing report on the abundance of fish and the industrious settlers. A treatise on overseas ventures, published in London three months later, noted that Guy's 'plantacon is very honest peacefull and hopefull and very lykelye to be profytable.'

Guy returned to England in the fall of 1611 with prospects of opening new plantations at Renews, south of St. John's, and on Trinity Bay. He returned to his colony in the spring of 1612, accompanied by sixteen women, presumably wives of the artisans, and was joined later by six apprentices. By year's end there were sixty-two colonists in the plantation.

Following two years of good fortune and mild winters,

the colony was plagued by problems in 1612. Guy's expansion plans had to be curtailed because of danger posed by pirates based at nearby Harbour Grace. Conflicts also arose between the colonists and fishermen from Devon and Cornwall, whose supporters in England denounced Guy's colony in the House of Commons. The severe winter of 1612–13 caused considerable illness, eight deaths, and the loss of most of the livestock, and investors soon became disillusioned with the colony's meagre returns.

One bright note was struck on 27 March 1613, when a weather log incidentally made note of the first-recorded white child born in Newfoundland. The father was Nicholas Guie, whose relationship to John Guy is unknown.

Two days after the child's birth, John Guy departed for London, never to return. He was elected mayor of Bristol in 1618 and represented the city for two terms in the House of Commons before his death in 1629. The colony was governed from 1615 to 1621 by Capt. John Mason, who during that period made the first English chart based on his hydrographic survey of eastern Newfoundland.

It was long assumed by historians that the colony at Cupids Cove was abandoned sometime after 1621, but archaeological evidence has revealed continuous occupancy, with a sharp population decline after 1750. The population of Cupids rose to 840 in 1836, and peaked at 1,424 in 1884. The current population is 868.

In 1910, on the tercentenary of Cupids's founding, a monument near the site of the original stockade was erected by the Newfoundland Historical Society. The City of Bristol donated a plaque for the monument, and the Old Colony Association (Newfoundland expatriates in Toronto) raised a 7- by 11-metre Union Jack – then believed to be the second-largest in the Empire – on a 41-metre pole half a kilometre east of the monument. Seventy-five years later, a new and larger Union Jack was donated by H.B. Dawe, owner of the local fish-packing plant; it is flown on holidays and other special occasions. Seeing the Union Jack fly in the stiff offshore breeze in 1991, I imagined the same scene in the days of the Sea-forest Plantation, and marvelled at the fortitude of the colonists.

Since the early 1970s, the Cupids Historical Society and the municipality have urged the federal government to build a modest museum or interpretation centre in Cupids, without success. The 400th anniversary of Cupids's founding is in 2010.

Acadia: A Land without Boundaries

To some, Acadia is a remote corner of New Brunswick, to others, a historic area of Nova Scotia's Annapolis Valley. Many believe it includes the three Maritime provinces and even extends into Québec. Although it is not shown on modern maps, Acadia occupies a special place in the hearts and minds of more than a quarter of a million people who proudly proclaim their Acadian heritage with a flag, an anthem, and a national day.

There is confusion about Acadia's origin and meaning, and few people agree on the exact area it encompasses. Yet it is a name of extensive geographical, historical, and cultural significance.

The name of Acadia can be traced to the discoveries of Giovanni de Verrazzano, who explored the coast of North America in 1524. He was struck by the beauty of Chesapeake Bay, calling it Archadia, a variation of Arcadia, the pastoral paradise of the Greek and Roman classics.

If Verrazzano produced a map of his voyage, none has survived. However, the places he named during his travels are shown on a 1527 map by Vesconte de Maggiolo, although Arcadia does not appear on it. The variation, Larcadia, first appeared on a 1548 map by Giacomo Gastaldi. He located it near what we now call Cape Cod. In the 1560s, another Italian map shifted Larcadia to the northeast, where it supplanted Tierra de los Bretons. In 1575, French historian André Thevet changed the name to Arcadie.

In 1599, King Henry IV of France appointed Pierre Chauvin de Tonnetuit lieutenant-general of Canada, the coasts of 'Lacadie,' and other areas of New France for ten years. This commission was transferred to Pierre du Gua de Monts four years later. De Monts used Lacadie in his petition to the king to undertake the exploration and colonization of New France.

A 1601 map by French cartographer Guillaume Levasseur used the name Coste de Cadie for what is now Maine. Two years later, Samuel de Champlain wrote a report entitled *Des Sauvages* in which he used Arcadie for the area we now call the Maritimes. In the early years of the seventeenth century, the form 1'Acadie became common in documents and on maps referring to Port-Royal and other settlements around the Bay of Fundy.

From the late 1620s to 1763, Acadia was an undefined territory east of New England and southeast of New France. The inhabitants of French extraction, cut off from other French colonies and with little immigration from their homeland, developed a distinctive Acadian culture. After their forced expulsion to the Thirteen Colonies and Europe in 1755, and the subsequent return of many to the region, their Acadian heritage was reaffirmed.

After 1763, the name Acadia was superseded by Nova Scotia, which applied to New Brunswick and peninsular

Nova Scotia as we know them today. Later, Cape Breton Island, Prince Edward Island, the southern Gaspé coast, and even Îles de la Madeleine were considered parts of Acadia.

Recent writers, such as Naomi Griffiths and Jean-Claude Dupont, often acknowledge the European evidence of Acadia's origins, but then proceed to suggest the name is derived from a native name for their lands, or from the Mi'kmaq expression for 'plenty,' 'fertile,' or 'camping place.'

The first reference to a possible Mi'kmaq origin may have been made by Thomas Chandler Haliburton, a Nova Scotia writer, judge, and politician, in his A *General Description of Nova Scotia*, published in 1823. In 1849, Nova Scotia scientist Abraham Gesner stated in *The Industrial Resources of Nova Scotia* that 'the terms Cadie and l'Acadie have evidently been derived from the Mi'kmaq *ākăde* – a place.'

The first protest against the Aboriginal origin for Acadia was lodged in 1896 by the renowned New Brunswick naturalist William F. Ganong, in the *Transactions* of the Royal Society of Canada. But twenty years later, he was disappointed to learn that few scholars agreed with him.

After 1917, most historians and writers acknowledged the European source as being correct, although historical geographer Andrew Clark wrote in 1968: 'Perhaps the most sensible conclusion is that the cartographic ancestry of Arcadie for various parts of the coast of eastern and northeastern North America prepared the way for the acceptance of "Cadie", "La Cadie", "L'Acadie" and so "Acadie" from its Indian source.' I believe the converse is true: place names ending with the Mi'kmaq expression *a-kaa'-di-k*, meaning 'occurrence place,' were spelled with the suffix *–acadie* by early explorers and map makers, reflecting the region's name of European origin. A few examples of such renditions are Chebenacadie (for Shubenacadie River, N.S.) and Tracadie (for Tracadie Bay, P.E.I.), shown on a French map drawn in 1744.

Acadians have contributed substantially to the Cajun culture of Louisiana, where there is a parish called Acadia. Acadia National Park is in the state of Maine, east of the mouth of the Penobscot River. As well, Acadia and Acadie have survived in numerous place names, including Acadieville, N.B., L'Acadie, west of Saint-Jean-sur-Richelieu, Qué., and New Acadia, P.E.I. Acadia University in Wolfville, N.S., received its charter as Acadia College in 1841. From 1924 to 1966, Acadia was a federal electoral riding in central Alberta; it probably received its name from the little community of Acadia Valley, 150 kilometres northeast of Medicine Hat, where Nova Scotians had settled in 1910. In 1990, the federal riding of Gloucester, N.B., was changed to Acadie–Bathurst to reflect its culture. The riding embraces the Acadian Peninsula, where more than 60,000 Acadians live.

Some contemporary writers have suggested that New Brunswick, where the vast majority of Acadians live today, is the centre of modern Acadia, and have looked to Moncton and its university as its focal point. But the soul of Acadia resides in many small towns and villages, such as Caraquet and Cap-Pelé in New Brunswick; Urbainville in Prince Edward Island; and Pubnico, Comeauville, and Chéticamp in Nova Scotia.

Many may regret that Acadia is not a country, with borders and legal institutions. But as the successor of the fabled beautiful land of the classics, perhaps it is best that modern Acadia resides in the hearts and souls of a people with a proud social and cultural heritage based on nearly four centuries of settlement in Eastern Canada. No matter where Acadians dwell, their rich literature, language, and music will tell them they are in Acadia.

Montréal: A People's City beneath a King's Mountain

When Jacques Cartier ascended 'the great river of Hochelaga and the route to Canada' in 1535, he arrived at the rapids below the site of the present city of Montréal on 2 October and was greeted by some 1,000 Iroquois, about one-quarter of the estimated population of the region. The next day, with proper decorum, he entered Hochelaga, a palisaded village of some fifty buildings. He then climbed the 223-metre, tree-clad mountain rising behind the village and named it either directly in honour of the king, François I, or indirectly as a feature worthy of the king.

Most scholars believe the name Montréal is a variation of Mont Royal, but few attempt to explain how and when the variation evolved. Some note that one of Cartier's companions on the 1535 voyage was the son of the Seigneur of Montréal in Languedoc, France, and suggest Cartier chose the name in honour of his family. Others connect the name with Cardinal de Médicis, Bishop of Montreale in Sicily. Still others look to a number of places with similar names in France and Spain, noting that they all have a location overlooking the countryside and concluding that Cartier gave the name as a generic term for a place with a distinctive mountain.

Jean Poirier, formerly with the Commission de toponymie du Québec, has thoroughly investigated Mont Royal's and Montréal's linguistic and toponymic roots, and has concluded that Montréal is derived from Cartier's name for the mountain.

The earliest reference to Montréal is found in a 1575 atlas by François de Belleforest, where the text, in a clear reference to the Iroquois village of Hochelaga, notes the 'ville les Chrestiens appelerent Montréal.'

The evolution of Montreal from Mont Royal has two

explanations. First, in sixteenth-century France, the words *royal* and *real* had the same meaning. For example, Cartier named a Newfoundland headland C. Royal in 1534; on a 1546 French map by Pierre Desceliers, it appears as *C. Real*.

Second, atlas maker Belleforest relied on a 1565 Italian translation of Cartier's writings. In it a plan of Hochelaga identifies the mountain as Monte Real. The reason for the use of *real* in place of the Italian *reale* is unknown, but may have been inspired by a Spanish or Portuguese map or document. As Poirier contends, it is easy to conclude that Belleforest was only writing another form of Mont Royal.

In 1612, Samuel de Champlain referred to the mountain as *montreal*. Twenty years later, he wrote of the Isle de Mont-real. In 1637, the *Jesuit Relations* referred to 'la grande Isle nommée de Mont-Real.'

In May 1642, Paul de Chomedey de Maisonneuve and a party of thirty-nine men, two priests, and nurse Jeanne Mance (founder of Montréal's first hospital) arrived at Hochelaga to establish the missionary colony of Ville-Marie and convert the Aboriginal people to Christianity. A year earlier, New France's Governor Montmagny had warned him of the danger of attack by Iroquois, but Maisonneuve would not be turned back, insisting that he had been instructed by the Société de Notre-Dame de Montréal in France to start a settlement there, and he would proceed even if all the trees on the island were to change into Iroquois!

Maisonneuve wrote of the colony of 'Ville marie en l'isle de Montréal en la nouvelle-France' in 1642, but Ville-Marie was superseded by Montréal in the early 1700s. As early as 1684, military officer and author Baron Lahontan wrote that 'this city is called Ville-Marie or Montréal.' In 1703, he signed a letter simply 'À Montréal.' In 1709, Jesuit Father Camille de Rochemonteix wrote that 'this city is situated on an island and has two names, which are Montreal and Ville-marie, the one by popular usage, the other by the priests of the seminary located there. The first name, Mont-royal, describes a large mountain in the central part of the island; by corruption, the mountain, the island and the city were called Montreal.'

Montréal was incorporated as a city by royal proclamation on 7 May 1792. Two hundred years later, the Montréal Urban Community has a population over 1.75 million. The city of Montréal, one of twenty-nine divisions of the urban community, has a population of 1,016,375. Montréal-Nord, Montréal-Ouest, and Montréal-Est have populations of 81,580, 5,250, and 3,525, respectively. The Town of Mount Royal, on the north side of the mountain, has a population of 18,500 in an area of less than 8 square kilometres.

'My Canada includes Québec' is a popular appeal from the four corners of Canada to the people of Québec to continue contributing their culture and passion for life to this great land. Because Montréal has endured for 350 years as Canada's spiritual centre, my Canada must continue to embrace Montréal.

Meech Lake – of Accord Fame – and Other Lakes Nearby

Nestled in the forested hills of Gatineau Park, less than a half-hour's drive northwest of the Parliament Buildings, is a chain of three lakes. Two have achieved national prominence: Meech Lake because it is beside the Thomas L. Willson House where the 30 April 1987 constitutional accord was signed; Harrington Lake because the house at its southern end became the prime minister's country retreat in 1959.

Meech Lake gets its name from Asa Meech (1775–1849), a Congregational minister in New England from 1800 to 1821 before moving his family to Hull Township in Québec. Three years after settling at the lake, he was granted 200 acres (80 hectares) and began clearing a farm. As well, Meech continued to minister in the area of Aylmer, Hull, and Chelsea, and gained a reputation for intellectual ability and spiritual influence. His original house, extensively remodelled over the years, has been preserved by the National Capital Commission (NCC).

On an 1870 plan made by Eugène Taché, the lake is called 'L. Charité,' and is said to have been named either for a French settler, François Lacharité, or for an Irishman whose name may have been Lacharity. Asa Meech's name was also attached to the plan, but for some reason it was spelled Meach, and this misspelling was confirmed by the Geographic Board of Canada in 1931.

In 1951, Marion Meech, a descendant of Asa, and some others asked the Canadian Board on Geographical Names to respect the correct spelling of Meech. The board, basing its decision on the long-established use of 'Meach' on maps, decided to maintain this version.

In 1982, the NCC sent the Commission de toponymie du Québec a copy of Asa Meech's will and other documents noting the spelling of Meech's name. The provincial commission was persuaded, and corrected the name to Lac Meech. Although it was thought that it would take some time to convince the public to make the switch, the change was almost immediately accepted by map makers and by the media. And, of course, publicity about the Meech Lake Accord has thoroughly imprinted the name (and correct spelling) in the minds of people all across Canada.

Harrington Lake, which is immediately northwest of Meech Lake, was a bird sanctuary and then the private estate of Col. Cameron M. Edwards prior to its acquisition by the Federal District Commission (predecessor of the NCC) in 1951. The name would appear to be a modified form of the name of John Hetherington, a son-in-law of Asa Meech. Hetherington, who was raised nearby, is thought to have built a house near the outlet (southeast end) of the lake at some time before 1850, although he is not mentioned in the 1861 census for the area. His brother, Joseph, lived nearby at the head (northwest end) of Meech Lake.

Eardley Township, in which the lake is located, was surveyed in 1850, and 'Harrington's Lake' is shown on the plan. The first grant in the township was given in 1860 at the head of the lake to James and David MacLaren, who sold the property to Louis Mousseau seven years later.

Although official plans and maps up to 1951 showed the name Harrington Lake, and this was used by local people, visitors, and residents of Ottawa and Hull were more familiar with Mousseau Lake or Lac Mousseau. Even Colonel Edwards called it Lac Mousseau. Letters to the *Ottawa Journal* and to the federal names board resulted in a review of the name of the lake. Arguing that the name of the first settler had priority, and long-established usage on maps should take precedence, the board decided to retain the name Harrington Lake.

In 1959, when the Prime Minister's Residence Act was passed, the form Harrington Lake (Lac Mousseau) was used in the English text, and the reverse in the French. This prompted the names authority in Québec to adopt Lac Mousseau (Lac Harrington) in 1962. This double form was subsequently shown on topographical and NCC maps.

Fearing that Harrington would disappear from maps, the Historical Society of the Gatineau in 1969 urged the adoption of Harrington Creek for the drainage of Lac Philippe into Lac Mousseau, and from there into Meech Lake. But the Commission de toponymie du Québec approved Ruisseau Mousseau, stating that the name Lac Mousseau (Harrington) would sufficiently honour the memory of the first settler. The society then asked for the approval of Harrington Ridge for the height of land west of the lake, but no action was taken on this proposal.

In 1987, the Commission de toponymie du Québec published a new gazetteer and omitted all mention of the name Harrington, even as a cross-reference to Lac Mousseau. Questioned about this, the commission stated that it regarded bracketed forms as transitory, and such forms were eliminated in the 1987 gazetteer. Thus, Harrington Lake no longer has official status for the lake, and although Lac Mousseau will likely attain wide use, it is not expected that the old name will be forgotten because of its association with the prime minister's country retreat.

The third lake, Lac Philippe, is one of the most popular in the national capital area for swimming, canoeing, and camping. It is on the 1850 plan of Eardley Township as Philip's Lake, and the form Philips Lake occurs on later maps. However, the Geographic Board of Canada endorsed the form Lac Philippe in 1931, and this is universally used by both English and French speakers. The lake may be named for either John Philippe, who had a farm nearby according to the 1861 census, or another farmer named James Philippe, who lived a few kilometres to the north.

While Lac Philippe and Meech Lake are open to the public, Harrington Lake/Lac Mousseau is the prime minister's private domain.

Ottawa: A Capital Name from a Fur-Trading People

In late January 1858, the 10,000 residents of the city of Ottawa received heart-warming news: on the last day of 1857, Queen Victoria had signed the despatch making Ottawa the capital of the United Province of Canada. Choosing Ottawa over five other applicants – Québec City, Montréal, Toronto, Kingston, and Hamilton – was greeted with much dissent by several politicians, but in the fall of 1865, the government was moved to Ottawa, the following year the Parliament Buildings were officially opened on the prominent hill overlooking the Ottawa River, and in 1867 the city was affirmed as the new capital of the Dominion of Canada.

As early as 1835 the rising lumber village of Bytown had been proposed as a potential capital for the projected union of Upper and Lower Canada. Kingston was designated the capital in 1841, but after the death of Governor General Sir Charles Bagot two years later, and his replacement by Charles Theophilus Metcalfe, 1st Baron Metcalfe, the capital was moved to Montréal, and after 1849 rotated between Toronto and Québec City for sixteen years. To select a permanent site for the capital, Governor General Sir Edmund Walker Head proposed in 1856 that the contending cities submit statements to the queen on the appropriateness of their place being chosen. Privately Head had decided that there was 'a choice of evils, and the least evil, I think will be ... in placing the seat of Government at Ottawa.' He himself spent much of 1857 in London persuading Colonial Secretary Henry Labouchere to obtain the queen's endorsement of his choice.

From 1826 to 1832, Bytown was the headquarters for the Rideau Canal project, which was constructed under the direction of Col. John By. It was legally incorporated as a town in 1850. During its first twenty-five years Bytown had acquired a reputation of being a rude, boisterious, vulgar rough-and-tumble lumbering and hard-drinking community. The town council in 1853 decided that it could be a credible candidate for the province's capital only if it discarded its bland and unpretentious name, and chose a more sophisticated title.

It just so happened that 1854 was an important anniversary: 200 years earlier, a crew of about 120 Aboriginals brought a small flotilla of furs from the west side of Lake Michigan to trade with the French at Montréal. They were subsequently identified as the Ottawas by the British, and *les Outaouais* by the French. To celebrate the anniversary, the 1853 Bytown council approved the renaming of the town as Ottawa. On the first day of January 1855, the city of Ottawa came into being.

The name's spelling followed an English rendering of the river's name, given to it in the late 1600s by French missionaries, explorers, and map makers. In 1613, Samuel de Champlain had referred to the river as *Rivière du Nord* and *Rivière des Algommequins*, after the Aboriginals he found during his voyage up the river to the present area of Pembroke. Récollet Brother Gabriel Sagard, who travelled in the valley in 1623–4, called it *Grande Rivière*, a translation of the Algonquin *Kitchisippi*. Both *Grande Rivière* and *Grand River* continued in common use well into the first half of the twentieth century. In 1680 French map maker Abbé Claude Bernou identified the river with the Ottawa nation by calling it *Rivière des Outaouais ou des Hurons ou des prairies*. Bernou's spelling followed that used by Intendant Jean Talon ten years earlier. The English form has its roots in Governor Marquis de Denonville's rendering of *Otawas* for the nation in 1687. In the mid-1700s, map maker John Mitchell portrayed *Riv. of the Outaouais or Utawas* on a map of North America. It was only in the early 1800s that the form *Ottawa River* became fixed in Canadian surveys and on maps.

The Ottawa nation was encountered in 1615 at the mouth of the French River by Champlain, who referred to them as *Cheueux Releues* (an archaic form of cheveux relevés), meaning 'standing hair,' in reference to the practice of tying their hair in a knot on top of their head. On meeting the Huron nation in 1623 in the area of present-day Simcoe County, Brother Sagard learned that these Iroquoian-speaking people called that Algonquian tribe of Ottawas living at the mouth of the French River, along the north shore of Lake Huron and on Manitoulin Island, *Andatahoüats*. In the 1640s and '50s the Jesuit missionaries referred them as *Outaouan, Ondataouaouat,* and other variants, including *Outaouac.*

The lives of the Huron and *Ondataouaouat* nations were dramatically transformed in 1649, when the Iroquois drove them from their territories. The *Ondataouaouat* found refuge in the area of Green Bay, on the west side of Lake Michigan. In 1653, after the threat of the Iroquois had abated, they brought a few pelts to Trois-Rivières to find out whether the French would be interested in more furs. Being assured the pelts were some of the finest ever seen, the *Ondataouaouat* returned the following year with a larger flotilla of furs. They continued the practice for several years thereafter, demanding that all trade with the French be controlled by them. Their name even became the general term for any Aboriginals carrying trade goods on the river.

The word *adawe* or *atawa* means 'to trade' or 'to buy and sell' in several Algonquian languages, including Algonquin, Ojibwa, and Cree. Several authorities have therefore concluded that the name *Ondataouaouat,* and subsequently *Outaouais* and *Ottawa,* were all derived from a phrase meaning 'to trade.' Other researchers have claimed *Ondataouaouat* may have something to do with either pendulous ears, or ears cut short, because of the similarity of the word for ear, *otawagama,* but such a conclusion is only speculative. Missionary Frederic Baraga suggested in 1878 that the name appeared to come from *watawawininiwok,* 'the men of the bull rushes,' but he probably just drew a parallel between two similar words. The tribe now calls itself Odawa.

The name of the capital city of Canada, therefore, has its roots in the name of an Algonquian tribe, identified in the early 1600s as *Ondataouaouat,* whose trading of furs after 1654 with the French resulted in the names *Rivière des Outaouais* and *Ottawa River.*

In December 1999, Ontario's Parliament created one of Canada's largest cities in area (2,696 sq. km.) by amalgamating the cities of Ottawa, Nepean, Gloucester, Kanata, Cumberland, and Vanier, the village of Rockcliffe Park, and the municipal townships of Osgoode, Rideau, Goulbourn, and West Carleton, creating the new city of Ottawa, with a population of over 742,000. The law establishing the new city came into effect on 1 January 2001.

A group of islands in Hudson Bay are known as the Ottawa Islands, with three islands bearing the names of prominent lumber barons in the national capital: Bronson, Booth, and Gilmour. A tributary of the North Branch Renous River in New Brunswick was named Little Ottawa Branch, possibly by Ottawa Valley lumberman Daniel McLachlin.

After the middle of the 1600s, a branch of the Odawa remained in Michigan, with some of their members migrating south and west to Pennsylvania, Illinois, Kansas, and Oklahoma. There are four counties called Ottawa in the United States, one each in Kansas, Michigan, Ohio, and Illinois. Three cities in the United States are known as Ottawa, one each in Illinois, Kansas, and Ohio, where they are all county administrative centres.

Toronto Has a Great Fish Tale to Tell

Toronto was first mentioned at its present site in the *Journals* (1765) of Maj. Robert Rogers, the noted commander of army rangers in the 1700s, as 'a proper place for a factory.' In 1760, Rogers had seen the remains of Fort Toronto, destroyed by the retreating French, as the British captured their North American possessions. He also noted that a branch of 'river Toronto' provided easy communication with Lake Huron (i.e., the Georgian Bay part of the lake).

The French had built three posts between 1686 and 1740, with the third being located at the present site of the Canadian National Exhibition. It was named for Antoine-Louis Rouillé, the Minister of Marine and Colonies, and continued until 1759. The second fort, built about 1720, at the mouth of *Rivière Taronto* (now Humber River), was called Fort Toronto, which was also the informal name given to Fort Rouillé. It was abandoned in 1730. Several eighteenth-century French and British maps identified it as Fort Toronto. The fort's name is spelled Toronto in all surviving records.

The river itself took its name from the voyageur route from Lake Simcoe to Lake Ontario, referred to as early as 1686 by New France's Governor General Marquis de Denonville as *Le passage de Taronto*. The route in turn received its name from a French name for Lake Simcoe, *Lac de Taronto*, which first occurs on a map dated to about 1680 by the National Archives, and attributed to French-court official Abbé Claude Bernou.

Finally, the lake received its name from the Mohawk

tkaronto, from the practice of the Huron and other Aboriginals of driving stakes into the water to create fish weirs in The Narrows, where Lake Simcoe empties into Lake Couchiching. The catching of fish there by means of weirs was such a significant activity among the Aboriginals that the Severn River, which drains Lake Couchiching into Georgian Bay, was also identified on maps as *R Taronto*.

Radiocarbon dating in the 1970s of surviving stakes revealed the weirs had been in continuous use for over 4,000 years. In 1615, Samuel de Champlain described the weirs as blocking the channels, with a few openings to collect fish in nets. In 1982, The Narrows was declared a national historic site. Because the stakes have been considerably threatened in recent years by power boating, anchoring, fishing, dredging, and bridge contruction, the Canadian Parks Service is considering The Narrows for designation as a World Heritage Site.

Huron-language expert John Steckley and several Mohawk speakers have interpreted *tkaronto* as 'where there are trees standing in the water,' which figuratively describes the stakes that formed the weirs in The Narrows.

Lake Simcoe is labelled *Le lac Ouentara* on a map titled *Novvelle France*, dated to 1641 by historical geographer Conrad Heidenreich. Variations of this Huron name occur on several maps during the following fifty years, and emerge again on maps during the period 1746–85 as *Oentaronk* and *Oentaronck*. Some writers have assumed this name provides the roots for Toronto, but Steckley states the linguistic elements of *Oentaronk* could not evolve as Toronto, although he notes it has an identical meaning to the Mohawk name.

Many French maps from the 1690s to the 1760s identify Lake Simcoe as *Lac Taronto*. On Vincenzo Coronelli's

1688 map of the Great Lakes area the name *L Taronto* appears above the words *Les Piquets*, 'the stakes.' The earliest spelling of *Toronto* for the lake is on a 1695 map by Coronelli. The use of the spelling *Toronto* on British maps, such as Herman Moll's of 1720 and John Mitchell's of 1755, may have been due to Baron de Lahontan, whose popular writings in 1703 and later about his travels around the Great Lakes were widely read. In the 1700s, the French name for the lake became Lac aux Claies, which translates as 'fish-weir lake.'

In his *Toronto: Past and Present* (1884), historian Henry Scadding interpreted Récollet missionary Gabriel Sagard's 1632 definition of the Huron *toronton – il y en a beaucoup*, 'there is much' – as suggesting a gathering of tribes, or 'place of meeting.' Many references repeat this definition, including *The Macmillan Book of Canadian Place Names* (1978) by William Hamilton; Nick and Helma Mika's *Places in Ontario*, vol. III (1983); my own notes under 'Place-names' in *The Canadian Encyclopedia* (1985 and 1988); and historian J.M.S. Careless's comments under 'Toronto' in the same reference. Professor Careless was on the right track when he suggested a preference for 'trees in the water' in his *Toronto to 1918* (1984).

In his earlier *Toronto of Old* (1878), Scadding had dismissed an origin from Sagard's *toronton* with the sense of 'plenty.' However, some writers and historians continue to repeat it. Historian William Kilbourn, perhaps influenced by Eric Hounsom's comments in his *Toronto in 1910* (1970), promoted the idea in his *Toronto Remembered* (1984), noting it was an appropriate concept for the Huron's land of plenty. 'So when anyone asks what Toronto means' Kilbourn wrote, 'I would suggest that the best reply is "abundance."'

In 1787, Governor General Lord Dorchester, finding the name Toronto in use at its present site, arranged what was called the Toronto Purchase from the Mississauga, embracing more than 1,000 square kilometres in the area of the city of Toronto and York Region. The following year surveyor Alexander Aitkin made a plan of the Toronto townsite, and Captain Gother Mann drew the *Plan of Torento Harbour.*

In 1792, Lieutenant-Governor John Graves Simcoe located the capital of Upper Canada at Niagara, which he called Newark. Judging the capital susceptible to attack from the United States, Simcoe proposed locating it at the Forks of the Thames, the site of present-day London. Lord Dorchester vetoed that location, but accepted Simcoe's selection of Toronto.

Disliking Aboriginal names, Simcoe introduced several English names in Upper Canada, including the renaming of Lac aux Claies for his father, Captain John Simcoe, who had served with Wolfe in the capture of Québec. On learning of a victory by the Duke of York in Flanders, Simcoe decided on 26 August 1793 that an English stamp should be impressed upon the new capital, and changed its name to honour the duke, George III's second son, Frederick Augustus (1762–1827).

British traveller Isaac Weld, who visited York in 1796, lamented the loss of sonorous Aboriginal names in his *Travels* (1799), writing that 'Newark, Kingston, York are poor substitutes for the original names of the respective places Niagara, Cataraqui, Toronto.' As early as 1804, after Simcoe's return to his beloved England, a petition was submitted to the legislature to reinstate Toronto. Because York could be confused with New York and other Yorks, and because Muddy York was a demeaning reference to the province's capital, the older name was restored on 6 March 1834.

Toronto has been the name of several other locations along the north shore of Lake Ontario. In late 1805, Alexander Grant, the administrator of Upper Canada – he served between lieutenant-governors Peter Hunter and Francis Gore – named Toronto Township at the mouth of the Credit River, 12 kilometres west of the Humber. In 1817, a post office at the mouth of the Ganaraska River, 95 kilometres east of Ontario's capital, was called Toronto. Three years later, it was changed to Port Hope, with Toronto being given then to a new office where Dundas Street crosses the Credit River in present-day Peel Region. Although a town called Toronto was laid out there in 1830, it was subsequently called Springfield, and, by 1900, renamed Erindale. In 1828, the Toronto post office was moved 5 kilometres northeast to Cooksville. When York was changed to Toronto in 1834, its post office was called Toronto City until 1837. After Cooksville was substituted for the other Toronto, 'City' was dropped from the capital's name.

Toronto Township was a municipality until 1967 when it became the town of Mississauga. Seven years later it was upgraded to a city. Toronto Gore Township, to the northeast of Toronto Township, was formed in 1819. In 1974, it was amalgamated with the city of Brampton. Just west of the mouth of the Humber, a post office called New Toronto was opened in 1892. It became an incorporated town in 1913. When Metropolitan Toronto was reorganized in 1967, New Toronto was amalgamated with the city of Etobicoke. Since 1998 it has been a neighbourhood in the city of Toronto.

The historical, cartographic, and linguistic evidence appears reasonably conclusive that the name Toronto is derived from the Mohawk description of the fish weirs 125 kilometres north of the city, with the name migrating by way of *Lac de Taronto, Passage de Taronto, Rivière Taronto*, and finally *Fort Toronto*. All other theories – 'place of meeting,' 'plenty,' 'harbour,' 'lake opening,' an Italian engineer called Tarento, trees on the Toronto Islands – lack crediblity.

On 1 January 1998, Metropolitan Toronto became the single city of Toronto. The cities of Etobicoke, North York, Scarborough, and York, and the borough of East York, were amalgamated with it, giving the new city an area of 638 square kilometres, and a population of 2,385,421.

Flin Flon: A Strange 'Professor' on Even Stranger Ventures

If names can be said to express a character of their own, surely the name Flin Flon evokes a feeling of adventure and vitality. The very sound of the name conjures up images of rugged rocks and wild waters.

On a winter's day in January 1975, so the story goes, Tom Creighton, a prospector and woodsman, went hunting for moose along the Saskatchewan–Manitoba border near the 55th parallel. He fell through the ice of a lake and on reaching shore he and his companions started a fire to help him dry out. The fire's heat melted a hole and through it they glimpsed a glint of gold. They proceeded to stake out claims.

This tale may be more fanciful than real. The truth may be that Creighton and his party were led to the deposit by David Collins, a Métis trapper and amateur prospector.

In the summer of 1915, Creighton and his associates – Dan Milligan, Dan and Jack Mosher, and Leon and Isador Dion – went back to work on their claim. When Jack Mosher brought some gold out of a hole, Creighton exclaimed: 'That must be the hole where old Flin Flon came up and shook his whiskers, so what do you say we call the discovery Flin Flon?'

And who was Flin Flon? He was the hero of an adventure novel, *The Sunless City,* a dog-eared copy of which had been found the previous year by Creighton on a portage between the Churchill River and Lac la Ronge in Saskatchewan. The book had been read and reread, although the last pages were missing, leaving a mystery for our northern travellers as to the fate of the novel's hero.

The principal character in the tale was blessed with the unlikely moniker, Professor Josiah Flintabbatey Flonatin, or Flin Flon for short. His strange adventure to the bottom of a lake in the Rocky Mountains in a submarine was written by Joyce Emerson Preston Muddock, a male writer of adventure stories, and published by F.V. White of London in 1905. The paperbound edition, which Creighton likely had, was printed in 1911.

The professor, a member of a so-called Society for the Exploration of Unexplored Regions, discovered a strange city abounding in gold. On learning that the city was ruled by strong women with men as their servants, he escaped back to the surface through an inactive volcano.

In 1932, Jack Carr, head of research for Hudson Bay Mining & Smelting, tried to secure a copy of the novel. After writing some 850 letters, he finally obtained a copy from a London bookdealer for a few shillings. In 1942, George Cole, Manitoba's director of mines, reported that only three copies were known to exist: Carr's copy (in the National Archives of Canada since the 1960s), another with a second official of the mining company, and a third in the British Museum.

Flin Flon, Man., a city of nearly 7,500 people, is located just east of the Saskatchewan–Manitoba boundary. A small part of the city, with a population of 330, is in Saskatchewan, where the boundary makes a 2.5-kilometre adjustment to the west. Nearby, to the southwest, is Creighton, Sask., a town of about 1,700 people, founded in

1948 and named for Tom Creighton, who died in 1949 aged 75.

In 1962, on the fiftieth anniversary of the extension of Manitoba to the 60th parallel of latitude (the boundary between the western provinces and the territories), a prominent statue of the adventurous Flintabbatey Flonatin, designed by cartoonist Al Capp in his noted Dogpatch style, was unveiled at the entrance to the city of Flin Flon.

Lake Laberge (Rhymes with Marge) and Other Yukon Names

The harshness of an unforgiving northland combined with frontier hardship and adventure are reflected in the many ballads and poems of Robert Service. Many young Canadians are nurtured on the verses Service wrote during the early years of the twentieth century, and acquire a yearning to discover more about the people who were lured by the spell of the Yukon, as revealed in these lines:

> The Northern Lights have seen queer sights,
> But the queerest they ever did see
> Was the night on the marge of Lake Lebarge
> I cremated Sam McGee.

In the 1860s, the Western Union Telegraph Co. engaged American naturalist William H. Dall to explore a route for a line in the Yukon River valley. The first large expansion of the river below Whitehorse Rapids he named for fellow explorer 'Michael Lebarge.' Neither is believed to have seen the lake, but heard of it from local Aboriginal people. The United States Geographic Board adopted the name Lake Lebarge, as well as other names in the Yukon given by Dall and Lt. Frederick Schwatka, and this, incidentally, was what prompted Canada to set up its own names board in 1898.

The Canadian authorities consulted church officials in 'Lebarge's' home town of Chateauguay, Qué., and received a reply from the explorer himself pointing out that his correct name was Michel Laberge. The lake's name has been officially spelled 'Laberge' since 1898, but its pronunciation still rhymes with 'marge.'

Service's poems broadcast the name Yukon to the world. As a name, it is simply the Gwitch'in word for 'the greatest river.' The river was named Youcon in 1846 by the Hudson's Bay Co. trader John Bell. He reached it in present-day Alaska, where he encountered some Gwitch'in, after travelling from Fort McPherson on the Mackenzie River. The Gwitch'in, who also lived on the lower Mackenzie River, did not occupy lands in the Yukon River's watershed within the Yukon.

In the area of Lake Laberge the river was known in the 1800s as Lewes River, named by Robert Campbell in 1843 for John Lee Lewes, then the Hudson's Bay Company chief factor at Fort Simpson on the Mackenzie River. In 1883, Lt. Schwatka substituted Yukon for Lewes. It was not until 1949, however, that the Canadian names authority dropped the name Lewes which, until then, was the official name for that portion of the Yukon River upstream from its juncture with the Pelly River.

The name Yukon was assigned to a district of the Northwest Territories in 1895; the district became a separate territory in 1898, with its seat of government at Dawson City. The capital was moved to Whitehorse in 1951.

In several of his poems Service included the name Klondike, a name synonymous with gold fever. His ballad 'The Trail of Ninety-Eight' mentions several other Yukon names. It ends with the lines:

> There were the tents of Dawson, there the scar of the slide;
> Swiftly we poled o'er the shallows, swiftly leapt o'er the side.
> Fires fringed the mouth of Bonanza; sunset gilded the dome;
> The test of the trail was over – thank God, thank God, we were Home!

The early miners had difficulty pronouncing 'Thronduik,' the Han name for the tributary river that joins the Yukon at Dawson City, and took to calling it Klondike. The Aboriginal word is said to describe the practice of the Gwitch'in of driving stakes into the riverbed to trap migrating salmon.

Rabbit Creek was renamed Bonanza Creek in 1896 after the discovery of gold there by George Carmack, Skookum Jim Mason, and Tagish Charlie.

After the gold rush the prominent hill overlooking Dawson City became known as Midnight Dome. Many people of Dawson City climb the hill on 21 June to watch, on or about midnight, as the sun barely sets before rising again.

In 1968, Robert Service's name was given to two features northeast of Dawson City. Mount Robert Service and Robert Service Creek will leave Service's name indelibly imprinted on the Canadian landscape.

Observing Selected Names in Particular Regions

Captain Vancouver's Legacy

In June 1792, British navigator Capt. George Vancouver sailed into Burrard Inlet and observed the forested setting where Canada's third-largest city stands today. Vancouver was almost fifteen months into one of the longest hydrographic surveys ever undertaken – four years, eight months, and twenty-nine days. His instructions were to survey the northwest coast of North America – from Mexico's Baja California to Alaska's Cook Inlet – to determine if a passage joined the Atlantic to the Pacific. But there was another purpose to his voyage: to ensure British control over part of the northwest coast.

Spain had tried to control the entire area from California to Alaska during the 1780s, while British traders were building a lucrative fur trade with Asia. In 1789, the Spanish seized two British merchant ships in Nootka Sound, claiming they were trespassing, and Britain prepared for a war. Realizing it could not mount a successful campaign, Spain agreed to the Nootka Convention in October 1790, forcing it to abandon its right to sole occupation of the area and to return the merchant ships to Britain.

Vancouver, then commander of the sloop *Discovery,* set sail from England on 1 April 1791, accompanied by the armed tender *Chatham.* Thirteen months later he sailed into what he called the 'supposed Strait of Juan de Fuca,' casting doubt on the truth of de Fuca's 1592 discovery there of a great passage into a broad sea with many islands. He spent May exploring Puget Sound, naming it and nearby Mount Baker, in Washington state, for two of his officers. Then he went ashore on King George III's birthday and named 'The Gulph of Georgia' – changed by Capt. George Richards in 1809 to the Strait of Georgia – in his honour. This incomprehensible change did not affect the name Gulf Islands, where many independent and creative people have chosen to live over the years.

But when Vancouver met Spanish explorers Dionisio Alcalá Galiano and Cayetano Valdés in late June, he was dismayed to learn that another Spaniard had already named the gulf Gran Canal de Nuestra Señora del Rosario la Marinera. Although Vancouver favoured recognizing names given by previous explorers, in this case he shifted the Spanish name to a narrow strait nearby so Britain could lay claim to the gulf. In 1859, Capt. Richards named the small strait after Capt. Alexandro Malaspina, commander of the Spanish settlement on Nootka Sound in 1791, and leader of Spanish exploration in the North Pacific.

Vancouver, Galiano, and Valdés agreed to work together in surveying the waters northwest of the gulf, with Vancouver and his crew concentrating on the the long, sinuous fiords penetrating the Coast Mountains. They spent the summer of 1732 charting the heads of Jervis, Bute, Knight, and Rivers inlets and the many intricate channels and narrows between Vancouver Island and the mainland. James Johnstone, the *Chatham*'s master, discovered that the strait Vancouver had named for him led to the Pacific. And northwest of Johnstone Strait, Lieut. William Broughton, commander of the *Chatham*, charted the waters around a group of islands which Vancouver called Broughton Archipelago.

In late August 1792, after exploring as far north as Burke Channel near Bella Coola, Vancouver sailed south to Nootka Sound to negotiate the land transfer from the Spanish to the British. But despite a cordial greeting by the Spanish commander Juan Francisco de la Bodega y Quadra, the two could not agree on the terms of transfer, and referred the matter back to their respective governments. Regardless, the two became friends, and Bodega (addressed incorrectly as Quadra by Vancouver) recommended that a prominent feature be jointly named after them. Having discovered that Nootka was on a huge island, Vancouver called it Quadra and Vancouver's Island, but the name Quadra soon fell into disuse. In 1903, the Geographic Board assigned the name Quadra to another island between Vancouver Island and the mainland.

After wintering in the Hawaiian Islands, Vancouver returned to the northwest coast in May 1793 to survey the many canals, channels, inlets, and sounds from Burke Channel to Revillagigedo Island near the southern tip of the Alaskan Panhandle.

While most of the names Vancouver gave in 1792 and 1793 honoured his officers and prominent men in the British Admiralty and navy, he also chose the surnames of noted political leaders for several features: Pitt, Hawkesbury, and Dundas islands; Grenville Channel; Portland Canal.

After another winter in the Hawaiian Islands, Vancouver resumed his survey in the spring of 1794, exploring the coastline from Alaska's Cook Inlet south to Point Conclusion, on Baranof Island in the Alaskan Panhandle.

On completing his survey, and convinced there was no passage between the Pacific and the Atlantic or an inland sea, Vancouver mapped the west coast of the Queen Charlotte Islands before returning to Nootka Sound. With no further instructions from Britain regarding negotiations with the Spanish, Vancouver sailed home.

By the 1860s, some seventy years after Vancouver explored Burrard Inlet, two towns – Gastown and Hastings – had been founded on its south shore. Then, in 1870, residents of Gastown renamed it Granville in honour of the British colonial secretary. The city that evolved might have remained Granville if it had not been chosen as the terminus for Canadian Pacific's transcontinental line in 1881, and if CP's general manager had not been William Van Horne. Van Horne, like Vancouver, was of Dutch ancestry, which led him to declare in 1885: 'This is destined to be a great city, perhaps the greatest in Canada. We must see that it has a name that will designate its place on the map of Canada. Vancouver it shall be, if I have the ultimate decision.' On 6 April 1886, Vancouver was formally incorporated as a city.

Several features along the west coast also honour the eighteenth-century navigator. In 1860, Capt. George Richards named Vancouver Bay in Jervis Inlet, where Vancouver stayed overnight on 17 June 1792. Vancouver River flows into the bay from the east. There is another Vancouver Bay at the entrance of Bute Inlet. And Vancouver Rock in Milbanke Sound, which Vancouver described in his journal as a 'very dangerous sunken rock,' was named in 1866 by British hydrographic surveyor Capt. Daniel Pender. Mount Vancouver, a peak of 4,828 metres near the Alaska–Yukon border, was named in 1874 by William H. Dall, scientific director of the Western Union Telegraph expedition.

Vancouver's meticulous survey helped maintain Britain's claim to part of the coast and ensure the eventual expansion of Canada to the Pacific. Along the way he also enriched Canada's toponymy: the names he gave and the many Spanish names he preserved on his charts were widely disseminated on Arrowsmith maps and British Admiralty charts. The only criticism of his naming practices is the almost total disregard of Aboriginal names, although he and his officers frequently had friendly encounters with the Indigenous people.

Mackenzie Expeditions Left a Trail of Names

During the summer of 1789, a young Scottish fur trader and explorer reached the shores of the Arctic Ocean after an epic journey down Canada's longest river. Four years later he led a party of ten men on an arduous journey over the Rockies and the Coast Mountains to the Pacific. These exploits at the end of the eighteenth century should be well known and the fur trader extolled as one of the greatest explorers of all time. But most Canadians know very little about the adventures of Sir Alexander Mackenzie.

Mackenzie was born in 1764 in Stornoway on the Isle of Lewis in Scotland. Ten years later his widowed father, Kenneth, emigrated with his young family to New York, where his brother, Murdoch, had a successful business. The father remained loyal to the Crown during the Revolutionary War, becoming an officer in the King's Royal Regiment. During his absence, Alexander was cared for by two of his father's sisters in Johnstown, N.Y. In 1778, they sent him to school in Montréal.

The next year, Mackenzie, then fifteen, was hired by John Gregory, a Montréal fur trader. Five years later he became a partner in the fur-trading firm of Gregory,

McLeod & Company, which later was absorbed by the North West Company.

In 1787, the North West Company sent Mackenzie to assist Peter Pond, who was in charge of one of the company's posts on the Athabasca River. Pond, an enigmatic and irascible figure from New England, had arrived in the western interior in 1775 and explored the country around Lake Athabasca (then called Lake of the Hills) and Great Slave Lake. Pond had been told by the Chipewyan that the lakes drained north to the 'Ice Sea.' However, when he learned that Capt. James Cook's discoveries on the West Coast suggested a large river draining from the interior, he compiled maps implying a link between Great Slave Lake and 'Cook's River.' Were there such a connection, it might have been a practical fur-trading route to the Pacific.

In 1788, Mackenzie succeeded Pond at the Athabasca post, and his cousin Roderick Mackenzie spent the year moving the headquarters to Fort Chipewyan on the south shore of Lake Athabasca, east of the mouth of the Athabasca River. (Fort Chipewyan was subsequently moved again in 1804 to its present site on the lake's northern shore.) There he planned a mission to determine if Lake Athabasca and Great Slave Lake did drain ultimately into the Pacific.

On 3 June 1789, he set out by canoe, accompanied by four French-Canadian voyageurs, a young German named John Steinbruck, a Chipewyan called English Chief, and a company of wives and retainers. In just forty days, Mackenzie's party reached the Arctic Ocean. From Great Slave Lake to the river's mouth they averaged 120 kilometres a day, one day canoeing an incredible 180 kilometres. Such distances were possible because the river flowed swiftly and had few treacherous sections. Moreover, the company was accustomed to canoeing twenty hours a day, sometimes starting as early as 2:30 a.m. Mackenzie was back at Fort Chipewyan on 12 September after an expedition covering nearly 4,800 kilometres.

Although Mackenzie used the name Grand River – perhaps from the Slavey *Deh Cho* – he derisively referred to it as 'River Disappointment' because it led only to the 'Hyperborean Sea.' When Aaron Arrowsmith made the map for Mackenzie's book *Voyages from Montreal* (1801), the name 'Mac Kenzies River' was shown twice, and Mackenzie, in his own observations on the geography of North America, referred to it as 'Mackenzie's River.' In writing to President Thomas Jefferson in 1807, Joel Barlow noted that the world had 'justly given the name Mackenzie to the great river of the north for the obvious reason, the merit of discovery.' However, when explorer John Franklin canoed the river in 1825 he found the name Great River in general use and urged the universal adoption of Mackenzie River in honour of its eminent discoverer.

During his first voyage of discovery in the Arctic in 1821, Franklin named a small point near the mouth of the Coppermine River for Alexander Mackenzie. Franklin had called it Cape Mackenzie, but in 1965, because the feature was judged by the Canadian Permanent Committee on Geographical Names to be too small to merit the word *cape*, it was altered to Mackenzie Point.

Mackenzie is used in five other names in the Northwest Territories. Mackenzie Bay at the mouth of the river was named officially in 1910, although it appeared on earlier maps. The river flows into the bay through a complex of braided channels of the Mackenzie Delta. This name was used by geologist Hugh Bostock in 1947 and officially approved the following year.

The District of Mackenzie was one of the three statutory divisions of the Northwest Territories, established along with the districts of Keewatin and Franklin in 1895. The districts no longer serve any administrative function but have been maintained on official maps because they remain in the official description of the Northwest Territories.

While Mackenzie remained curious about the land beyond the mountains to the west of the Mackenzie River, he wrote about them simply as 'the Mountains.' John Franklin in the 1820s and Kaye Lamb, who edited *The Journals and Letters of Sir Alexander Mackenzie* (1970), referred to them as the Rocky Mountains, but the Rockies are not now considered as extending into the Northwest Territories. The system of mountains and ranges from the Liard River in the south to the Ramparts River in the north, and west to the Yukon border, is called the Mackenzie Mountains. They were named officially in 1939 for Alexander Mackenzie – not the explorer but the prime minister of Canada from 1873 to 1878, who, by coincidence, was born in Dunkeld, Scotland, in 1822, the place where the explorer had died suddenly two years earlier while on his way from Edinburgh to his estate at Avoch in Ross-shire.

The Mackenzie Highway was built as an all-weather road in the 1940s from Grimshaw, Alta., to Hay River, N.W.T. Later it was extended to Fort Simpson and Wrigley.

After his journey to the Arctic, Mackenzie, determined to become proficient with surveying instruments, went to London in 1791 to learn more about both navigation and surveying. He returned to Fort Chipewyan in 1792 and that same year travelled to Fort Fork on the Peace River, near the present town of Peace River, to make plans to find a practical route to the Pacific.

On 9 May 1793, now twenty-nine, he set out from Fort Fork accompanied by Alexander McKay, six voyageurs, and two Aboriginal hunters and interpreters. By 1 June he had reached a section of the Parsnip River in present British Columbia. The District Municipality of Mackenzie was created at that site in 1966 as the logging and mining centre for that part of the Rocky Mountain Trench. The preplanned town now has a population of 5,800.

The most arduous part of Mackenzie's trip was the portage through rough terrain from the head of the Parsnip River to James Creek, Herrick Creek, and then McGregor River, a tributary of the Fraser. (He assumed the Fraser was the Columbia River; Simon Fraser proved otherwise fifteen years later.) Perhaps at the summit – the lakes here are appropriately called Arctic Lake and Pacific Lake – Mackenzie may have been able to see several snow-covered peaks of the Rockies some 100 kilometres to the southeast. In 1916, one of them was named Mount Sir Alexander in his honour. The 3,277-metre mountain is 75 kilometres north of McBride.

By 22 June, Mackenzie had followed the Fraser to a point halfway between the present cities of Quesnel and Williams Lake. Twenty-eight years later the North West Company established Fort Alexandria at this site, naming it for the explorer. The present community of Alexandria has a population of 82.

On the advice of Aboriginals, Mackenzie decided against proceeding farther south on the turbulent Fraser. Instead, he turned back to a river he called the West Road – today it is officially called West Road (Blackwater) River – and worked his way west along an Aboriginal track towards the Pacific. By 17 July, he reached the highest point of his travels. Mackenzie Pass, 1,830 metres, is the summit between the valleys of the Dean River to the north and Bella Coola River to the south. Mackenzie Valley is north of the pass in the valley of Kohasganko Creek, and Mount Mackenzie is to the west. These three names were given by the Department of National

Defence while surveying the area in 1953. When Mackenzie had trekked over the pass he saw a 'stupendous mountain' to the south; the official name now is Stupendous Mountain.

Mackenzie reached Bella Coola on 19 July 1793, and continued canoeing to the west until he came to a rock in Dean Channel where he took astronomical observations to determine his latitude and longitude. On the rock he inscribed: 'Alexander Mackenzie, from Canada, by land, the twenty-second of July, one thousand seven hundred and ninety-three.' In 1926, the British Columbia government named the site of the rock Sir Alexander Mackenzie Provincial Park. Mackenzie then returned to Fort Chipewyan, taking only thirty-three days to cover the circuitous 2,000-kilometre route via the Fraser, the Parsnip, and the Peace. Remarkably, on his two journeys to the Arctic and the Pacific, no one accompanying him died or suffered serious injury.

After his two prodigious exploratory trips, Mackenzie spent much of the next fifteen years in Montréal, where he remained active in the fur trade. In 1801, on one of his many trips to London, his *Voyages from Montreal* was published. The book created immense interest in Britain and the United States, and was translated into French, German, and Russian. In 1802, at age 38, he was knighted. After 1808, he remained in Britain, where he married a distant cousin and had three children.

With eleven features named for Sir Alexander Mackenzie in the Northwest Territories and British Columbia, including Canada's longest river, it may well be argued that Mackenzie has been adequately honoured. But other than in the names of schools in Edmonton and Inuvik, the explorer does not appear to have been remembered in the names of the nation's institutions. In July 1989, 200 years after Mackenzie reached the Arctic, a plaque was unveiled at his Scottish gravesite. Four years later, the words he inscribed near Bella Coola were carved on a granite slab at his grave.

'88 Olympics: Calgary, Kananaskis, and Mount Allan

Holding the 1988 Winter Olympics in Canada vaulted Calgary onto the global scene so that, like Innsbruck or Grenoble, Lillehammer or Nagano, its name is easily mentioned around the world without reference to region or country.

Calgary was named in 1876 on the suggestion of Col. James F. Macleod, who was assistant commissioner of the North West Mounted Police when a post had been established the previous year where the Elbow River joins the Bow. He chose the name to honour a small place with family associations and happy memories on the Island of Mull in Scotland. As adduced by the noted historian George F.G. Stanley, the name in Scotland most likely means 'bay farm' in Scots Gaelic. Other suggested origins, such as 'clear running water' (the meaning advanced by Macleod) or 'cold enclosure' (a meaning promoted by writers Eric J. and Patricia M. Holmgren), would appear to have no etymological basis.

Bow River, thc principal tributary of the South Saskatchewan River, rises in the Rocky Mountains northwest of Banff. It derives its name from the Aboriginal practice of gathering saplings in the valley to make bows. An Arrowsmith map of 1802 shows Bow Hills, and another Arrowsmith map of 1822 refers to 'Bow or Askow River.' The French word for 'bow' is reflected in the name Lac des Arcs, some 90 kilometres upriver from Calgary at Exshaw. It was named in 1858 by Eugène Bourgeau, the botanist on the Palliser Expedition.

Kananaskis River, a major tributary of the Bow, rises about 60 kilometres south of Exshaw in some of the most awe-inspiring mountain landscape in North America. John Palliser named the river in 1858 for a Cree who, according to legend, miraculously recovered from the blow of an axe. In this area, widely known as Kananaskis Country, and 12 kilometres south of Exshaw, is Mount Allan, chosen in 1984 for the '88 Olympic alpine ski events. This 2,800-metre peak was named in 1948 for John Andrew Allan (1884–1955), who had headed the department of geology at the University of Alberta from 1912 to 1949, and had undertaken extensive geological studies in the Rocky Mountains. A 1987 study undertaken for the Stoney Chiniki Band Council revealed the names Châse Tîda Baha ('burnt timber hill') and Wâtaga îpa ('grizzly hill point') for it. In 1985, the Alberta government chose the name Nakiska for the ski resort at Mount Allan. This name, derived from a Cree word meaning 'to meet,' refers to athletes from all over the world who came together in February 1988.

Immediately to the west of Mount Allan is Mount Lougheed, which may be the peak named Windy Mountain in 1858 by Bourgeau. In 1886, George M. Dawson recorded the name Wind Mountain for it, and this was made official in 1911 by the Geographic Board of Canada. In 1928, the board renamed the mountain for the late senator Sir James Lougheed (1854–1925), the grandfather of Peter Lougheed. In 1982, the Alberta Historic Sites Board confirmed the name Windtower for the northwesterly peak of the Mount Lougheed massif, and in 1985 restored the name Wind Mountain to its southeasterly peak. Kananaskis Provincial Park, at the head of the Kananaskis River, was renamed Peter Lougheed Provincial Park in 1986 by the provincial government.

Southwest of Mount Lougheed is Mount Sparrowhawk – the original choice for the Olympic downhill ski events – which was named in 1922 for a destroyer that had served in the Battle of Jutland during the First World War. To the south of it are several other named mountains that recall officers, cruisers, warships, and destroyers of the same battle: Glasgow, Chester, Engadine, Galatea, Indefatigable, Invincible, Warrior, Black Prince, Shark, Warspite. And in the same area adjacent to the British Columbia boundary are many other peaks that honour noted foreign leaders of the First World War: Joffre, Foch, Jellicoe, Putnik, Smuts, Leman, Hood, Brock; three senior British officers: Haig, Beatty, Evan-Thomas; and Canadian generals: Burstall, Currie, Mercer, Turner. Northeast of Mount Allan is Mount Lorette, named after a First World War battle at Lorette Ridge in France.

Straight south of Mount Allan is Mount Bogart, named for Donaldson Bogart Dowling (1858–1925), who explored this area in 1904 for the Geological Survey of Canada. Dowling named Mount Kidd, the next mountain to the southeast, for John Alfred Kidd, then a trading-post manager at Morley, the centre of the Stoney First Nation Reserve west of Calgary.

Across the Kananaskis River valley from Mount Allan is Fisher Range, named by Palliser in 1858 for George Fisher, a British astronomer and a family friend. A prominent elevation of the range, Mount McDougall, was named by George Dawson in 1884 for Rev. George McDougall (1821–1876), a Methodist missionary among the Stoney, and for his sons, David and Rev. John McDougall.

To the north of Mount Allan, and overlooking Exshaw, is Mount McGillivray, which was named in 1957 for Duncan McGillivray (*ca* 1764–1825), a partner of the North West Company, who travelled through the area in 1801. Just east of Mount McGillivray, a peak with a distinctive heart-shaped summit reveals the name given to it in 1957: Heart Mountain. Just west of Mount McGillivray is Pigeon Mountain, named by Bourgeau in 1858, and beyond it is one of the most photographed mountains in the Rockies: The Three Sisters, named by Dawson in 1886 for its three similar peaks. Finally, next in succession are Ha Ling Peak (until 1998, officially known as Chinamans Peak), named for a Chinese miner who had climbed it in 1896; and Mount Rundle, named by James Hector in 1859 for Rev. Robert Terrill Rundle, a Methodist missionary to the Cree in the 1840s.

Exshaw was named by Sir Sandford Fleming after a son-in-law, William Exshaw. Five kilometres upriver is Dead Man's Flats, which recalls the killing in 1904 of John Marrett by his brother, François. Canmore is a railway name given in the 1880s by Canadian Pacific, possibly for the Scottish King Malcolm III (d. 1093), known as Canmore.

The Thousand Islands: There Are Really 1,149

One of the most idyllic regions of North America embraces the Thousand Islands in the St. Lawrence River.

The early French explorers, on ascending the river to Fort Frontenac (now Kingston), were impressed with the great number of islands, and concluded they must number at least a thousand. The French historian Charlevoix noted in 1721 that 'we passed through the midst of a kind of archipelago which they called Mille Isles. I believe there are about 500.' A 1727 map by de Léry called them 'Les Mil Isles.'

Many of the travel brochures claim there are at least 1,800 islands. In the early 1800s, the British and American boundary surveyors determined that there were

1,692, deciding that any rock with at least one tree could be called an island.

In 1966, an official count by the Canadian Permanent Committee's names secretariat revealed 1,015 islands compiled on large-scale topographical maps from Amherst Island west of Kingston to McNair Island at Brockville, a distance of 80 kilometres.

A methodical check in 1985 of the large-scale navigation charts puts the number at 1,149 islands and islets. The number on the Canadian side is 665, with 241 officially named; on the American, 484, with 126 named.

Numerous rocks barely awash or rocks abutting larger islands probably account for the previously inflated numbers.

A Mississauga legend states that the islands resulted from an accident when a blanket holding Manatoana, or 'Garden of the Great Spirit,' was dropped by its carriers, breaking the garden into a thousand pieces.

Very few of the Thousand Islands had individual names prior to the War of 1812. The largest, Wolfe Island, known to the Iroquois as Ganounkousenot, and to the French as Grande Isle and Isle Buade, recalls the British general who led the capture of Québec in 1759. Other prominent British military leaders from that period honoured in the names of large islands include Gen. William Howe and Gen. Jeffery Amherst.

After the War of 1812, Capt. William Fitzwilliam Owen surveyed the St. Lawrence between Lake Ontario and Cornwall, and published a series of five Admiralty charts in 1828. In naming the islands he drew on many elements relating to British military engagements on the Great Lakes and in Upper Canada, as well as noted British military leaders.

Among gunboats on the Great Lakes during the war, with their names applied in the Lake Fleet Islands, are

Axeman, Aspasia, Belabourer, Bloodletter, Camelot, Deathdealer, Dumfounder, and Endymion.

The Navy Islands include the names of officers who distinguished themselves in engagements during the war. Among them are Downie Island (Cmdr. George Downie, killed on Lake Champlain in 1814); Hickey Island (Frederick Hickey, commander of the *Prince Regent* on Lake Ontario); Mulcaster Island (Capt. William Howe Mulcaster, who was severely wounded at Oswego, N.Y.); Spilsbury Island (Capt. Francis Spilsbury, who was involved in several engagements on Lake Ontario); and Owen Island (probably not for Capt. W.F., but for his brother, Sir Edward Owen, who commanded forces on both Lake Champlain and the Great Lakes).

Some of the Admiralty Islands off Gananoque were given names by Owen for prominent figures in the British Admiralty. Among them were islands for the Rt. Hon. Charles Philip Yorke, First Lord of the Admiralty in 1811, and Viscount Melville, the First Lord from 1812 to 1827. Possibly because these islands had local names prior to Owen's survey, his names were not accepted locally; Yorke is Bostwick Island; Melville is Hay Island.

A number of other names assigned by the Owen survey have recently been changed to conform to preferred usage in the area. Among them are Squaw Island (formerly Brock), Leek Island (Thwartway), Tremont Park Island (Tidds), and Sugar Island (St. Lawrence).

The well-known Grenadier Island honours the distinguished British Grenadier Regiment. Known to the Mississauga and the early travellers as Toniata, the name Grenadier occurs on a map relating to the Treaty of Ghent, 1814. Captain Owen called it Bathurst Island, probably for the third Earl Bathurst (1762–1834). Grenadier was given precedence in 1909.

Hill Island, crossed by the Thousand Islands (Ivy Lea)

Bridge to New York state, honours Gen. Rowland Hill. An 1874 survey by Charles Unwin identified it as Leroux Island, but the name assigned by Owen was confirmed in 1909.

Upriver from Brockville is the Brock Group, where several islands commemorate officers who served with Sir Isaac Brock in the War of 1812. Among them are Myers Island (Lt.-Col. Christopher Myers, commander at Kingston in 1813), Sheaffe Island (Gen. Sir Roger Sheaffe, who succeeded Brock at Kingston), De Watteville Island (Maj.-Gen. Abraham de Watteville), De Rottenburg Island (Maj.-Gen. Francis, Baron de Rottenburg), Cockburn Island (Admiral Sir George Cockburn), Conran Island (Maj.-Gen. Henry Conran), and Stovin Island (Maj.-Gen. Richard Stovin, commander at Montréal, 1812–14).

Many of the islands on the American side reflect a British connection, having also been named by Owen. One of the largest is Wellesley Island, which commemorates Arthur Wellesley, the Duke of Wellington. Others include Carleton Island (Sir Guy Carleton) and Picton Island (Sir Thomas Picton).

One of the most frequently visited American islands is Heart Island (formerly Hart), where George Boldt, owner of the Waldorf Astoria in New York, began but never completed a castle, and where his chef, who became the renowned Oscar of the Waldorf, concocted the famous Thousand Island Dressing.

In 1904, St. Lawrence Islands National Park was established to preserve some of the sensitive ecology of these fragments of the Canadian Shield. The park now includes part of the mainland at Mallorytown Landing, plus all or part of twenty-one islands and eighty-three rocky islets. Among the Thousand Islands, the complex network of channels includes the intricate Lovers Lane, the narrow Needle's Eye, the sharp-angled Fiddlers Elbow, the curiously named International Rift, and the carefree Wanderers Channel.

Some of the most pleasant-sounding names of Canada are scattered among the Thousand Islands. Included are Kitsymenie, Kalaria, Apohaqui, Minota, Arabella, Beaurivage, and Manomin.

Now, as for the Thirty Thousand Islands in Georgian Bay ... another essay or two perhaps.

Relocating the Lost Villages

The construction of the St. Lawrence Seaway in the 1950s was one of the most ambitious engineering projects ever undertaken in North America. The dams, canals, and powerhouse in the International Rapids section between Montréal and Lake Ontario had a profound impact on the region's physical and human geography. Some 8,000 hectares (20,000 acres) of farmland and all or part of nine villages on the Ontario side of the St. Lawrence River between Cornwall and Iroquois were inundated when Lake St. Lawrence was created. Two new towns, Long Sault and Ingleside, were established, partly of new homes and partly of houses moved from the flooded villages. Morrisburg's business district was moved and all of Iroquois was relocated.

Some of Ontario's oldest and most historic settlements – founded by United Empire Loyalists in the 1780s – were submerged under the waters of the man-made lake. Their names – Mille Roches, Moulinette, Dickinsons Landing, Wales, Woodlands, Farrans Point, and Aultsville – are remembered in the names of streets, roads, and islands, and in the Mille Roches, Moulinette, and Aultsville halls of Cornwall's St. Lawrence College.

Mille Roches ('thousand rocks'), 8 kilometres west of Cornwall, was adjacent to a rough and rocky section of the river's Long Sault Rapids. From the 1780s to the 1830s, it was a landing place for a brisk business in transferring freight and passengers. Its importance declined when the Cornwall Canal was built in 1836, but it remained a small service centre until the 1950s.

Moulinette, 3 kilometres farther west, was named for the small grist mill built by pioneer settler Adam Dixon, who also built a distillery, a foundry, an Anglican church, and a handsome residence.

Five kilometres southwest of Moulinette was the little community of Dickinsons Landing, where Barnabas Dickinson established a stagecoach line between Kingston and Montréal. When the Grand Trunk Railway was built from Montréal to Toronto in the 1850s, the stop north of the village was called Dickinson's Landing Station. In 1860, the Prince of Wales travelled by train to the station and returned to Montréal by the St. Lawrence River. In honour of his visit, the station was renamed Wales. It eventually grew into a small urban settlement.

Woodlands, a little community 4 kilometres west of Dickinsons Landing, was named because it was a firewood supply point for steamers on the St. Lawrence. Three kilometres farther west was the little urban centre of Farrans Point, where a Farrand family had lived in the early years of settlement. The last of the submerged group was Aultsville, named in 1854 for Samuel Ault, who served as an MLA and an MP from 1861 to 1872.

In 1954, the Ontario–St. Lawrence Development Commission was set up to transfer 6,500 displaced residents to the new townsites. The commission decided to build a new urban centre at the west side of Cornwall Township to receive people from Mille Roches and Moulinette, and another planned community 8 kilometres to the west in

Osnabruck Township for the residents of that township whose properties were to be inundated.

The Osnabruck Township council endorsed the name Sunnyvale for its new urban complex in July 1955, but two months later changed it to Avondale. Believing this might be confused with nearby Avonmore, and because there was another Avondale to the east in Charlottenburgh Township, several residents urged the adoption of another name, with Osnabruck, Wales, and Kanata being suggested as alternatives. A vote was held and Wales emerged as the preferred choice, but new opposition soon arose because only one of the former places was being commemorated in the new name. The township reeve, Thorold Lake, happened to drive by a house in Aultsville called Ingleside, and suggested this name to the council. This was accepted, and subsequently endorsed by the Canadian Board on Geographical Names, as the Geographical Names Board of Canada was then called.

In early 1956, the Mille Roches and Moulinette Chamber of Commerce proposed Longue Sault as the name for their new planned urban centre in Cornwall Township, and this was endorsed in July by the township council. The Cornwall *Standard–Freeholder* pointed out that the proper form was Long Sault, in keeping with the rules of the French language, and the Canadian Board on Geographical Names approved this form. However, angry residents claimed long usage of Longue Sault in the community, and mustered ample evidence for their claim. In February 1957, the newly named Longue Sault Chamber of Commerce pressed for reconsideration of the place's name, but to no avail.

Names for the twenty-three new islands that would be formed by the flooding were drawn up by the Ontario–St. Lawrence Development Commission in the spring of 1956. Names of communities and prominent persons in the valley's history provided inspiration and the names were endorsed by the Canadian names board. When Lionel Chevrier, then president of the St. Lawrence Seaway Authority, was advised of the new names, he proposed a number of changes, and several of his suggestions were adopted.

Wales, Mille Roches, and Moulinette are each recalled in the name of an island. Dickinson Island is for Dickinsons Landing. Woodlands Islands – East Woodlands Island, Centre Woodlands Island, and West Woodlands Island – all are named for Woodlands. Farrans Point is remembered in the name Farran Point, a peninsula that extends into the St. Lawrence beside Ingleside. Ault Island, the largest of the Canadian islands created by the flooding, is adjacent to the old site of Aultsville and just offshore from Upper Canada Village, where pioneer buildings and traditions are preserved.

Long Sault Rapids is recalled in the name of the urban centre as well as the Long Sault Parkway, which connects eleven islands in St. Lawrence Provincial Park to the mainland at Ingleside and Long Sault.

In the spring of 1958, James Duncan, chairman of the Hydro-Electric Power Commission of Ontario, proposed naming the new reservoir Lake St. Lawrence. It was approved by the Canadian names board in June, and later accepted by the United States Board on Geographic Names. Some newspapers, including the Cornwall *Standard-Freeholder,* objected to its blandness. (They also reported incorrectly that the official Canadian name was Lake Elizabeth.) Alternative suggestions flooded into newspapers and the Canadian names board; among the more bizarre ones were Lake Elizamerica, Lake Indedom (independence and dominion), Lake Canusa, Lake Uscan, and Lake Canarican, but the original proposal prevailed.

Memories of the submerged communities have been

kept alive by the Lost Villages Historical Society, founded in 1977, and its museum, opened in August 1992 at Ault Park, east of Long Sault. Inundation Day has been celebrated twice, first on 1 July 1983, and again on the same day in 1988. Many families were uprooted nearly forty years ago, but the great beauty of the islands and shoreline of St. Lawrence Provincial Park and the charm of Upper Canada Village have provided a large measure of spiritual compensation for the residents of the Seaway Valley.

Great Divide Passes in the Rockies

Some of the most historic place names in Canada are those of the mountain passes on the Great Divide in the Rocky Mountains.

From Akamina Pass near the United States border to Monkman Pass in north-central British Columbia, there are seventy officially named passes. North of Monkman Pass, the Great Divide skirts west of the Rocky Mountain Trench and Williston Lake, so that the Rockies in this area cease to be the continental divide between Pacific waters on the west and Atlantic (Hudson Bay) and Arctic waters on the east and north.

Three of the passes are crossed by both railways and highways: Crowsnest, Kicking Horse, and Yellowhead. A highway over the Great Divide uses Vermilion Pass.

A railway and a highway also go through Pine Pass in the Rockies, but at this point these mountains are not the continental divide.

Another railway was built in 1983–4 through an unnamed pass between Monkman Pass and Pine Pass to serve the Quintette and Bullmoose coal mines at Tumbler Ridge.

Two other passes became well known as routes during the fur-trade era: Athabasca and Howse. Monkman Pass has neither a railway nor a highway, although several efforts have been made to link Prince George with Dawson Creek and Grande Prairie by building a road through it.

The area of Crowsnest Pass (elev. 1,357 metres) was explored in 1858 by Thomas Blakiston of the Palliser Expedition and in 1882 by George Dawson of the Geological Survey. In the 1800s, the story was told of *ka-ka-iu-wut–ishis-tun* by the Cree and *ma-sto-eeas* by the Blackfoot, both literally meaning 'the raven's nest,' but translated as 'the crow's nest.' The hill where the Cree and Blockfoot believe ravens originally nest is thought to be 30 kilometres east of the present Crowsnest Mountain, with the latter having been given that name erroneously by white explorers. The two Aboriginal nations consider the area of the hill to be the 'crow's nest country,' accounting for the transfer of the name to the river and the pass.

Crowsnest Pass attained importance in 1897 when the Canadian Pacific Railway received a subsidy to build a railway through the pass and to develop the mineral deposits of southern British Columbia, for which it guaranteed fixed rates for the eastbound shipping of grain and flour. This Crowsnest Pass Agreement, extended in 1925 to all railways leading east to Lake Superior, kept the costs of grain transportation down, but resulted in poor development and maintenance of rail lines. In late 1983, a new 'Crow rate' was worked out to allow the costs of shipping grain to rise, although the railways are still supported by large federal subsidies. At the same time the railways agreed to spend $16 billion by 1992 for new rolling stock and improvement of service.

Vermilion Pass (elev. 1,651 metres) is 42 kilometres

west of Banff. It links the Bow River valley in Alberta with the Kootenay River valley in B.C. The name is derived from the red, yellow, and orange ochre beds in mineral springs 9 kilometres southwest of the pass. The springs were frequented through time immemorial by the Kootenay and the Blackfoot for decorative paint.

After Dr. James Hector crossed Vermilion Pass on horseback in 1858, he returned to the east side of the Great Divide by a tumultuous tributary of the Columbia River. Kicking Horse Pass (elev. 1,627 metres) received its fascinating name when Hector was kicked by his horse and knocked senseless. The river, by association, became the Kicking Horse River.

Although the grade from the top of the pass down what came to be known as 'the big hill' into Field, B.C. (elev. 1,241 metres), was greater than 4 per cent, the Kicking Horse was selected for the first railway route to Vancouver, the CPR completing its line through the pass in 1884. (The grade was reduced to 2 per cent with the building of the famous spiral tunnels in 1909.)

Otto Klotz, who undertook an astronomical survey along the CPR right-of-way in 1886, introduced the name Wapta for the river (from the Stoney word for 'river'), but the Hector name, Kicking Horse, prevailed. The Trans-Canada Highway, opened officially in 1962, traverses the pass.

Howse Pass (elev. 1,525 metres), occasionally touted in recent times as a practical highway route to shorten travel time between Edmonton and Vancouver, was crossed in 1810 by Joseph Howse, a Hudson's Bay Company trader. The pass had been crossed earlier in 1807 by David Thompson, who gave it the name Howse Pass on his 1814 map. Howse had been in charge of Carlton House, near present-day Prince Albert, Sask., from 1799 to 1809. He retired to England in 1815.

Athabasca Pass (elev. 1,748 metres) was crossed by David Thompson in 1811, and became thereafter an important route to the Columbia River system. The pass takes its name from the Athabasca River, which rises southeast of the pass; the name in Cree means 'where the reeds are,' a description of the marshy delta where the river enters Lake Athabasca some 1,000 kilometres away in northeastern Alberta.

At the summit of the pass is a small round lake called Committee Punch Bowl, where Sir George Simpson treated his companions to a bottle of wine in 1825. The name is a tribute to the governing committee of the Hudson's Bay Company.

Yellowhead Pass (elev. 1,133 metres) was not explored until the 1820s, but ultimately became the most important rail and highway pass north of the Kicking Horse Pass. The pass and the B.C. community of Tête Jaune Cache both honour a fur trader.

For many years the names of Jasper Hawes, François Decoigne, and Pierre Hatsinaton were advanced as the 'yellow headed' trader in question; but in 1984 David Smyth, a Parks Canada historian, established beyond doubt that the blond trader involved was really Pierre Bostonais. This mixed-blood Iroquois had worked for both the Hudson's Bay and the North West companies and was killed in the upper Peace River region in 1827.

The pass, called merely a portage during the fur-trade era, also became known as Leather Pass in the late 1800s, as hides had earlier been carried through it by traders from the Athabasca River on the east to New Caledonia on the west. Other names noted in the literature include Cowdung Pass, Leatherhead Pass, Jasper Pass, Jasper House Pass, Tête Jaune Pass, and Rocky Mountain Pass.

Sir Sandford Fleming failed in the 1880s to persuade the CPR to follow the route of the Yellowhead Pass, but it was ultimately chosen by both the Grand Trunk Pacific and the Canadian Northern. These were united in the 1920s into the Canadian National Railways, which has continued to use the pass as its main freight route to the Pacific coast. It is also used by Via Rail for passenger traffic between Vancouver, Edmonton, and Winnipeg. As well, the pass is the route of the Yellowhead Highway, which follows the same general route – Winnipeg, Saskatoon, Edmonton, Jasper, and on to the coast.

Monkman Pass (elev. 1,082 metres) was discovered by Alexander Monkman in the winter of 1921–2 while on a trapping and hunting trip. The pass was seriously considered for a railway to the Peace River country of Alberta in 1925.

A new provincial park recently established at the pass, and extensive coal developments north of the pass at Tumbler Ridge, may encourage the opening of a highway through the pass, a route long promoted in the Grande Prairie region.

The names of these seven passes through the Great Divide illustrate the diversity in geographical naming in Canada: The transfer of a Native name from a nearby feature – Athabasca; the translation of a description given by two Aboriginal nations – Crowsnest; the occurrence of a decorative paint – Vermilion; an incident – Kicking Horse; a mixed-blood Iroquois fur trader – Yellowhead; a Hudson's Bay trader – Howse; and a hunter and trapper from the region – Monkman.

Of course there are many other pass names (Deadman, Palliser, Assiniboine, Miette, Simpson, to name but a few), and each has its own interesting tale of discovery and exploration.

Commemorating Prominent Individuals and Honouring Certain Family Names

Sir Guy Carleton: Names Recall a Revered Governor

Ask yourself this question: What individual has had the greatest influence in shaping Canada, its federation of provinces and territories, and its unique blend of languages, traditions, and laws? Many people come to mind, but my choice above all is Sir Guy Carleton, Baron Dorchester, the predominant figure in the governing of the British Province of Quebec (spelled without an accent) for most of the years from 1766 to 1796.

At a time when victorious nations customarily imposed their laws, languages, and customs on the people of conquered territories, Carleton conceived the 1774 Quebec Act, which recognized French civil law, freedom of religion, and the right of Catholics to hold public office. That was more than a half-century before Britain emancipated Catholics. As commander-in-chief of British forces in the original Province of Quebec, he successfully defended the citadel of Québec City in 1775, and turned back the American invaders. Carleton participated in 1791 in the division of the Province of Quebec into the provinces of Upper and Lower Canada, with the

English language and civil law dominant in one, and the French language and civil law in the other.

Carleton played central roles in two cataclysmic events that inexorably determined Canada's future. A favourite of Gen. James Wolfe, he commanded one of the three battalions in 1759 that defeated the forces of Gen. Montcalm on the Plains of Abraham. He played a key role in transporting 30,000 British troops and 27,000 Loyalists and their families from New York in 1782 and 1783, and persuading the governments of the original provinces of Nova Scotia and Quebec to grant them free land and a year's provisions.

Guy Carleton was born in 1824 in Strabane, County Tyrone, Ireland. Before his eighteenth birthday he was commissioned an ensign in the British army, rising to the rank of lieutenant-colonel when he joined Wolfe in 1758 at the capture of Louisbourg. Although he had no civil government experience, his nomination in 1766 as lieutenant-governor of the old Province of Quebec was supported by senior officials in the colonial department and by King George III himself. He was made governor-in-chief in 1768, and continued in that office for ten years. In 1782, Carleton was appointed commander-in-chief in North America. Four years later, when he was created Baron Dorchester, he became commander-in-chief of British North America.

The earliest naming honour for Sir Guy Carleton occurred in Prince Edward Island when Samuel Holland, surveyor general of British North America, named the closest land to the mainland Carlton [*sic*] Point in 1765, and named two little bays Guy Cove and Carlton [*sic*] Cove. After the point became the site of a ferry terminal in 1917, it was renamed Borden Point for Prime Minister Robert Borden, and Guy Cove became Port Borden. In 1995 the municipal community of Carleton Siding, 3 kilometres inland, and the town of Borden were merged to form the municipal community of Borden–Carleton. The small community of North Carleton is 4 kilometres to the north.

In 1783, a detachment of disbanded soldiers and their families sailed from New York to Port Mouton in southwestern Nova Scotia, built 300 houses and called the place Guysborough to honour the heroic efforts of Sir Guy in managing the move. The name continued there as that of a township for many years thereafter, although the 800 settlers moved to Chedabucto Bay in eastern Nova Scotia in 1784 after a fire had completely destroyed their houses. They also transferred the same name to their new home, with the townsite of Guysborough being laid out in 1790, and Guysborough County being created in 1836.

In 1784, Col. Timothy Hierlihy and other officers of the Nova Scotia Volunteers obtained a grant of 21,600 acres in the area of present Antigonish County, and arranged for some Loyalists settlers to take up land there. On learning of Carleton becoming Lord Dorchester in 1786, Hierlihy named his projected townsite and township Dorchester in honour of him. However, the settlement failed, and by the early 1800s had been supplanted by Antigonish, 10 kilometres to the southwest.

Carleton is a small community in southwestern Nova Scotia, 25 kilometres northeast of Yarmouth. It was founded in the 1830s and first called Temperance. Later it was renamed after nearby Lake Carleton, named earlier for Sir Guy Carleton, but now known as Raynards Lake.

In adjoining Shelburne County is Carleton Village, 10 kilometres south of the town of Shelburne. A fort was built at the mouth of Shelburne Harbour in 1783–4 to protect the Loyalists settlers being transferred from New York to Shelburne. According to its postmaster in 1905, it was named for Thomas Carleton, Sir Guy's youngest brother, who was lieutenant-governor of New Brunswick from 1784 to 1817, although he had returned to England in 1803.

Regarded as the key figure among the founders of New Brunswick, Thomas Carleton was honoured in the names of Carleton Parish in Kent County in 1814, Carleton County in the Saint John River valley above Fredericton in 1831, and Mount Carleton, the province's highest mountain, given in 1899 by natural scientist William F. Ganong. Mount Carleton Provincial Park was created in 1985.

In 1783, when New Brunswick was still part of Nova Scotia, Governor John Parr gave the name Parr Town to the heart of the present-day city of Saint John, and the area of what is now known as Saint John West he called Carleton, for Sir Guy Carleton. It was still well known as that at the beginning of the twentieth century.

During three time periods, Carleton served for a total of sixteen years in Quebec, where he was strongly supported by the seigneurs and the clergy, it should not be surprising that many places were named for him in that province. He is described in *Noms et Lieux*, Québec's voluminous study of its geographical names published in 1994, as one of the most sympathetic Britons to the cause of the Québécois.

Carleton Township on the south side of the Péninsule de la Gaspésie was named about 1768 while Sir Guy was still lieutenant-governor of the Province of Quebec. The town that developed there was earlier known to the Acadians as Tracadièche (from the Mi'kmaq for 'place of many herons'), but about 1787 it was renamed Carleton by the Loyalists. In 1928, the 609-metre mountain north of the town was named Mont Carleton, a name in use locally for a number of years for what had also been called Mont Tracadigash.

Dorchester County in the Beauce country south of Québec City was named in 1792. Although the county is no longer part of the municipal structure of Québec, it is recalled in such names as Saint-Prosper-de-Dorchester and Sainte-Marguerite-de-Dorchester.

To strengthen the defence of the southern entrance to Lake Ontario, the British built a fort in 1778 on a small island south of Wolfe Island, and named it for Sir Guy. However, the military significance of Carleton Island declined after the signing the Treaty of Paris five years later. The British continued to occupy it until the War of 1812, although after the signing of Jay's Treaty of 1795, it had become American territory.

Carleton County at the junction of the Ottawa and Rideau rivers was named for Sir Guy in 1798, perhaps by Peter Russell, then administrator of the government of Upper Canada. In 1969 the county, with the addition of Cumberland Township, became the Regional Municipality of Ottawa–Carleton. As of 1 January 2001, the regional municipality disappeared and was replaced by the new city of Ottawa. Carleton retains a strong presence in the area in the name of Carleton University (founded as a college in 1942), with a student population of 14,800. The town of Carleton Place in nearby Lanark County was named in 1830, not for Sir Guy, but for Carlton Place, a street in central Glasgow.

Dorchester Township was created in 1798 east of London in Middlesex County, and named for Lord Dorchester. It was divided in 1851 into North Dorchester and South Dorchester townships, with the former remaining in Middlesex County, and the latter transferring to the new Elgin County. Dorchester is an urban centre in the Municipality of Thames Centre, created in 1999 through the amalgamation of North Dorchester and West Nissouri townships.

There are many streets in eastern Canadian cities called Carleton and Dorchester, but the best known is Dorchester Boulevard, renamed in 1985 by the city of Montréal to honour the founder of the Parti Québécois, René Lévesque, who had died that year. However, the boulevard extends into Westmount, where the name remains unchanged. The first bridge built across Rivière Saint-Charles in Québec City was opened in 1789, and named for Lord Dorchester. The crossing at the same site is still identified on maps as Pont Dorchester.

Fraser: A Thoroughly Geographic Name

Fraser is one of the most honoured surnames when it comes to naming geographical features in Canada. The most prominent feature is the Fraser River, the mighty drainage system of British Columbia.

The name Fraser has its roots in a Norman family whose armorial bearings included the blossom of the strawberry (*fraise*). The first Frasers arrived in Britain with William the Conqueror. They acquired lands in Scotland's East Lothian in the twelfth century, and subsequently spread throughout Scotland. Several Frasers distinguished themselves in military campaigns in Britain's defence of its American colonies in the 1700s, and in the fur trade and related exploratory travels.

One might expect only three or four dozen Fraser names in the Canadian Geographical Names Data Base, maintained by the Centre for Topographic Information in Ottawa. Surprisingly, there are more than four dozen currently used in British Columbia alone, while Québec has four dozen more, and Ontario and Nova Scotia each have more than three dozen. In total, there are 218 geographical features in Canada with Fraser as part of their names, another 15 names with Frazer, and a further 3 with Frazier. In addition, Fraser (or Frazer/Frazier) occurs as a part of 103 names previously used for features that now have other names.

The Fraser River was named by David Thompson in 1813, five years after Simon Fraser had made his historic journey to its mouth. Alexander Mackenzie had assumed in 1793 it was the Columbia, and Simon Fraser expected he would arrive at the Pacific well below the 49th parallel. Discouraged when he found it was not, he left the river unnamed.

The river is only one of many features in British Columbia named for the illustrious partner of the North West Company. Fraser Lake, 125 kilometres west of Prince George, was named by John Stuart, one of Fraser's clerks. The present village of Fraser Lake on the lake's south shore has a population of 1,300. Fraser's first name is honoured with Simon Bay on the north side of the lake. Fort Fraser, said to be the first settlement in the so-called Oregon Territory, was established in 1806 by Simon Fraser as a fur-trading post at the east end of the lake. The community of Fort Fraser now has a population of 450.

Near the source of the Fraser River in Mount Robson Provincial Park in the Rockies is Mount Fraser (elevation 3,270 metres). Its peaks are named Simon, Bennington, and McDonell in honour of Fraser's first name, his birthplace in Vermont, and his wife's maiden name. Bennington Glacier is on the British Columbia side of the Continental Divide; Simon Glacier and Fraser Glacier drain into Alberta's Simon Creek.

One of the stretches of the Princess Royal Channel between Princess Royal Island and the mainland, 90 kilometres south of Kitimat, is called Fraser Reach. It was named in 1866 for Donald Fraser, then a prominent lawyer and landowner in Victoria.

In Port Alexander, at the north end of Vancouver Island, is Fraser Island, named about 1860 by Capt. (later Admiral Sir) George Richards for Alexander Fraser Boxer, then the master of H.M.S. *Alert*.

Eighty kilometres northeast of Vernon is Bill Fraser Lake and Bill Fraser Creek, named for a widely admired guide who lived in the area from 1907 to 1963. Mount Fraser and Fraser Creek, 55 kilometres north of Kamloops, also are named for the same Bill Fraser. Fraser Island at the north end of Adams lake, 100 kilometres northeast of Kamloops, was named in 1972 for another Bill Fraser, a logger and forest-service patrolman living at Chase in the 1960s.

On the west side of the Fraser Canyon near North Bend is Fraser Peak, named in 1956 for Trooper William E. Fraser, who was killed in action during the Second World War. Near the British Columbia–Alaska boundary, 30 kilometres north of Stewart, is a 2,440-metre mountain called Mount White-Fraser, named for George R. White-Fraser, a member of the International Boundary Commission, 1904–11. (He served during the First World War in France, and died in 1920.)

Fraser Falls on the Stewart River in the Yukon is named for a member of an 1885 prospecting party. Fraser Creek, 125 kilometres southwest of Whitehorse, was named in 1902 for assistant surgeon S.M. Fraser of the North West Mounted Police, who was a mining recorder and customs officer at Dalton Post from 1901 to 1903.

In 1847, John Rae named Fraser Bay, on the west side of Melville Peninsula in Nunavut, for a friend who was a chief trader of the Hudson's Bay Company (likely Paul Fraser, who lived from 1797 to 1855). Six years later, Elisha Kane, an American explorer, named Cape Fraser on the east coast of Ellesmere Island for Prof. John F. Fraser of the University of Pennsylvania.

Where Hudson Bay and Hudson Strait meet, and off the northwestern tip of Nottingham Island, is Fraser Island. It was named in 1965 for Robert James Fraser (1887–1965), who served as the first Dominion Hydrographer from 1948 to 1952, and was the father of J. Keith Fraser, general manager and publisher of *Canadian Geographic*, 1982–90. Because Robert Fraser had contributed forty-five years to the hydrographic surveying of Canadian waters, the hydrographic service wanted to reward him with a perpetual memorial in the area of Hudson Bay, where he had undertaken the first hydrographic surveys during the years 1910–14. The Canadian Hydrographic Service first proposed a sea-floor feature, but I recommended selecting the then unnamed island so the name could be displayed on topographical maps.

Saskatchewan has four Fraser features named for Second World War casualties: Fraser Bay, in Reindeer Lake, for Gunner Arthur R. Fraser; Fraser Creek, which flows into the bay, for Pte. Robert W. Fraser; Fraser Lakes, west of Reindeer Lake, for Flight Sgt. John S. Fraser; and Fraser Lake, northwest of Lac la Ronge, for Flying Officer Alexander M. Fraser. Fraser Bay in Otter Lake, an expansion of Churchill River in Saskatchewan, was named in 1964 for a pioneer teacher and missionary, Canon Albert Fraser. Fraser Rapids, on Foster River north of Lac la Ronge, was named in 1966 for James Fraser and his wife who were homesteaders in the Pambrun area southeast of Swift Current.

Three features in Manitoba are named for Frasers killed in action during the Second World War: Fraser Island, in Gilchrist Lake east of Lake Winnipeg, is for Flying Officer John C. Fraser; Fraser Lake, near Lynn Lake, is for Flying Officer Alexander D. Fraser; and Fraser Creek, west of Churchill, is for Flight Sgt. Ian Fraser. Fraserwood, a small community 15 kilometres west of Gimli, was called Kreuzburg until 1918, when the name was changed to honour a local merchant, Samuel J. Wood, and his fiancée, Annie Bell Fraser.

Fraser Township, west of Pembroke, Ont., was named in 1854 for Col. Alexander Fraser of Fraserfield, Glengarry County, who was a member of the Legislative Council of Canada in 1841.

A post office was established in 1882 in the community of Fraserburg, in the present town of Bracebridge, with Alexander Fraser its first postmaster. The name recalls Fraserburgh, Scotland, founded by Sir Alexander Fraser in the late 1500s.

Fraserdale, 110 kilometres north of Cochrane on the Ontario Northland Railway, is named for Alan Fraser, a civil engineer on the railway when it was built north to Moosonee in 1930.

South of Lake Nipissing is Fraser Creek, which crosses land granted to John Burns Fraser in 1886.

Malcolm Fraser acquired the seigniory of Mount Murray, east of La Malbaie in Québec's regional municipal county of Charlevoix-Est, in 1761. He had served with distinction with Wolfe at Louisbourg and Québec, and was wounded in a skirmish at Ste-Foy in 1860. On Rivière Comporté, east of La Malbaie, is a set of falls named for him.

North of La Malbaie are two lakes, 18 kilometres apart, officially called Lac Fraser. The current 1:50,000 topographical map (1976) identifies each as Fraser Lake. This happened because of a federal mapping policy, in effect until 1976, that prescribed English generics for English names. I argued, unsuccessfully, that the Frasers of La Malbaie had been intermarried with the French from the early 1800s, and noted that Fraser itself was not an English name. The next edition of the map will portray Lac Fraser for the two features.

Alexander Fraser (1761–1837), a son of Malcolm Fraser, served with the North West Company from 1797 to 1809. He spent the winter of 1797 at Bedford House on Reindeer Lake in present-day Saskatchewan. An island in the lake is named for him. In 1802, he bought the seigniory of Rivière-du-Loup-en-Bas. In his honour the village of Fraserville was established in 1850, becoming a town in 1874. In 1919, the name was changed to Rivière-du-Loup. In the 1871 *Dominion Directory,* the following leaders of the community were noted: Edward Fraser, seignior and president of the Fraserville Institute; William Fraser, seignior and justice of the peace; and Alexander Fraser, advocate. Today, the town has both a Fraserville Street and a Fraser Street.

Several Fraser families from Scotland acquired lands in the Lake St. Francis region of the St. Lawrence River after 1818. In Québec's township of Dundee, James and Alexander Fraser developed farms adjacent to Pointe Fraser and Ruisseau Fraser. Across the lake on the Ontario side, Donald Fraser purchased an estate in the early 1800s at a site that became known as Fraser Point, and his descendants remained there until 1951. In front of the point are several small islands called crabs; one is known as Fraser Crab.

There are seven streams in New Brunswick called Fraser Brook. One or more of them may be named after Donald Fraser, who established sawmills at River de Chute in 1877 and in Edmundston in 1911. After 1917, Fraser Companies (now Fraser Inc.) set up pulp and paper mills in Edmundston and in Madawaska, Me.

Bishop William Fraser (1778–1851) was the first Roman Catholic bishop of Halifax, although he retained his residence in Arichat on Isle Madame, adjacent to Cape Breton Island. Four of his brothers – Colin, Thomas, John, and David – and a sister, Mrs. Donald Chisholm, established the community of Fraser's Grant, 20 kilometres east of Antigonish. Fifteen kilometres south of Antigonish on the South River, Alexander Fraser installed saw and grist mills in the mid-1800s. The post office there was called Fraser Mills from 1884 to 1937. Sunnybrae in Pictou County, 25 kilometres southeast of New Glasgow, also had a post office called Fraser Mills from 1848 to 1876. Fraser Settlement in the Musquodoboit Valley was the site of a grant given to another Alexander Fraser in 1813.

There are only two Fraser names in Newfoundland, both in Labrador. Fraser Lake, 100 kilometres north of Churchill Falls, may be named for James D. Fraser, well known at the turn of the century for his dog teams, which he used in fur trading on the Labrador coast. Fraser River, which flows into the Atlantic at Nain, may be named for James Fraser, who commanded Moravian vessels on the Labrador coast for some thirty-five years after 1782.

Prince Edward Island has only a single Fraser name, Mary Fraser Island in Malpeque Bay.

George Dawson: Inspiration for More than Twenty-five Names

The most brilliant and versatile scientist in Canada a century ago was arguably George Mercer Dawson. Best known as a geologist, he was also an accomplished botanist, zoologist, ethnologist, historian, geographer, linguist, and yes, even a toponymist. Add to that explorer, adventurer, diplomat, spellbinding speaker, and pastoral poet, and you have the quintessential Victorian-era gentleman and scholar – and a worthy inspiration for the names of more than twenty-five Canadian mountain peaks, inlets, islands, settlements, and animals.

The best known are Dawson Creek, B.C., and the Klondike Gold Rush capital of Dawson City, Yukon, which both commemorate Dawson's pioneering surveys of those regions. In two instances, Dawson himself gave his name to the features he encountered. Otherwise, other people named places after him, an indication of the esteem in which he was held, and of the nature of the man whose physical deformities and boundless energies earned him a number of admiring sobriquets, such as 'Little Giant' in a 1974 biography by Joyce Barkhouse – 'Little' because a childhood illness had left him head and shoulders shorter than an average man, with a hunched back and weak lungs. Despite these handicaps he never shied from arduous explorations.

Dawson was born in Pictou, N.S., in 1849, the son of Margaret Mercer and John William Dawson, a respected geologist and provincial superintendent of education. In 1855, the family moved to Montréal, where George's father became principal of McGill University and one of the first Canadian scientists of international note.

George Dawson studied at McGill and then graduated first in his class from the Royal School of Mines in London. In 1873, he was appointed naturalist and botanist with the British North America Boundary Commission, and spent two years studying the natural resources along the 49th parallel between Lake of the Woods and the Rocky Mountains. An island in Lake of the Woods he called Middle Island was renamed Dawson Island, possibly by Andrew Lawson, who undertook an extensive geological survey of the area in 1883. Nearby are Dawson Channel and Dawson Bay.

In 1875, Dawson joined the Geological Survey of Canada, established thirty-three years earlier to develop mineral industries in Canada. He became the Survey's assistant director in 1883 and director in 1895, a position he still held at the time of his early death in 1901, at age 52.

Dawson's first assignment with the Survey was a study of the geology of British Columbia. After completing extensive surveys of the southern part of the province, and recommending a route for the Canadian Pacific Railway, he went to the Queen Charlotte Islands in 1878, principally to evaluate the coal deposits there. He also took time to put together a report on the customs of the Haida, and to compile comparative vocabularies of the Haida and adjacent mainland tribes.

During his summer in the Queen Charlottes, Dawson fixed the present written forms of thirty Haida place names, and named another seventy-five features, usually using simple descriptives – Bag Harbour, Slug Rock – but also honouring noted scientists of the day, including Charles Lyell, Thomas Huxley, and Charles Darwin. He named Dawson Head, on a small island on the west side of Graham Island (the largest of the islands), for himself, and Rankine Islands, on the east side of Moresby Island, for his brother and assistant, Rankine Dawson.

In the years following his survey, Dawson was commemorated in the Queen Charlottes by other scientists and surveyors in the names of eight more features: Dawson Inlet and Dawson Harbour (1897), Dawson Islands (1910), Mercer Lake (1914), Dawson Point and Dawson Bay (1921), Mercer Point (1945), and Dawson Cove (1948). In addition, a rare caribou on the islands was identified in 1900 as *Rangifer tarandus dawsoni* by Ernest Thompson Seton, the noted naturalist and nature writer.

1n 1879, Dawson undertook an arduous reconnaissance from the mouth of the Skeena River, now the site of Prince Rupert, B.C., inland through the mountains and eventually to Edmonton. In the Peace River country near the Alberta–B.C. border, he assigned the name Dawson's

Brook to a minor tributary of the Pouce Coupé River, which flows into the Peace. According to Philip and Helen Akrigg, in their *British Columbia Place Names* (1997), Henry MacLeod, a CPR explorer who met Dawson in the area, may have changed the name to Dawson Creek – the name adopted when a village was established on the site in 1919. The settlement grew, particularly after the railway arrived in 1931, and again after 1942, when it became Mile 0 on the Alaska Highway. Incorporated as a city in 1958, Dawson Creek had a population of almost 11,100 in 1991.

Dawson's most remarkable overland journey was a 2,200-kilometre trip in the summer of 1887 on the Stikine, Liard, Pelly, and Yukon rivers in the Yukon and northern B.C. His carefully prepared map of the Yukon identified the Klondike River as a possible source of gold. When the gold rush brought thousands into the area ten years later, the trader and prospector Joseph Ladue asked William Ogilvie to survey a townsite at the river's mouth. Ogilvie said he would if he could name the town for the finest man he knew. On receiving news of the honour from Ogilvie, the 'Little Giant' pencilled in Dawson City on his map of the Yukon.

Ogilvie had honoured his estimable friend earlier, in 1888, with Mount Dawson, east of Lake Laberge. But Dawson asked two years later to have it called Mount Laurier after Wilfrid Laurier, then leader of the Liberal Party of Canada. Ogilvie may also have named Dawson Range, a chain of mountains southwest of the Yukon River.

In the Selkirk Mountains, east of Revelstoke, B.C., there is another Dawson Range. Its highest point, at 3,390 metres, is Mount Dawson. It was named in 1888 by William Green, a British alpinist and author of a book on the Selkirks.

Dawson Peaks are five summits of a mountain on the west side of Teslin Lake, northeast of Atlin, B.C. The name was noted by John Gwillim in his 1901 report on the Atlin mining district, and may have been given by Arthur St-Cyr, who surveyed the B.C.–Yukon boundary in 1899 and 1900.

Joseph Tyrrell, Dawson's assistant on an 1881 Alberta expedition, explored the area of Lake Winnipegosis in 1889, and named its northwesterly extremity Dawson Bay. Dawson Bay is also a small fishing community on the west side of the bay.

Tyrrell undertook several geological surveys in the Chesterfield Inlet, Thelon and Dubawnt river areas of present-day Nunavut. About 1893, he named Dawson Inlet, on the west coast of Hudson Bay between Arviat and Rankin Inlet.

A 1910 geological map of the Lake Nipigon area portrays several islands and bays with the names of scientists, including Logan, Murchison, and Humboldt. Dawson Island, in the centre of the lake, may have been named by Donaldson Dowling, who surveyed the area in 1894 with William McInnes, and on his own in 1898.

During a zoological survey of the Yukon River in the 1880s, Clinton Merriam, chief of the United States Biological Survey, encountered a new species of mouse and named it *Clethraonomys rutilus dawsoni.* I agree with Barkhouse, who says in her biography that the 'Little Giant' would be more pleased with this honour than any other, given the image of the little animals scurrying freely across the same wild lands he had explored and mapped.

Victoria: The Most Honoured Name in Canada

No individual has been more honoured than Queen Victoria in the names of Canada's public buildings, streets, populated places, and physical features. Her name appears more than 300 times on our maps. This is not surprising, given Victoria's long reign during the greatest glory of the British Empire, from 1839 to 1901.

Every Canadian can think of at least one public building or street named for her. Among the notable public institutions are Victoria University in Toronto, Royal Victoria Hospital in Montréal, and Victoria General Hospital in Halifax and in Winnipeg. Among Canada's 280 postal divisions, more than half have at least one thoroughfare identified by the name Victoria. Most were named to honour the Queen, but one that was not is Thunder Bay's Victoria Avenue, named for Victoria McVicar, a Lakehead landowner in the late 1800s, noted for her dealings with some celebrated builders of the West, among them Louis Riel, founder of Manitoba, and Sir William Van Horne, builder of the CPR.

The first honour for Princess Victoria may date from 1829. Pierre-Georges Roy wrote (1906) in a book on Québec's names that Pointe Victoria in the Saguenay had been named that year for her. But a check of mid-nineteenth-century charts and maps has failed to reveal the point so named.

In 1831, John Ross, while searching for the Northwest Passage, entered a small bay on the east side of Boothia Peninsula and named it Victoria Harbour for the thirteen-year-old princess. Subsequently, she formally granted him permission to use her name for this remote and minor water feature in Canada's Arctic. From then on, explorers, map makers, and administrators assigned the name Victoria to a multitude of lakes, points, islands, and other geographical features all over the Canadian map.

On the occasions of her golden (1887) and diamond (1897) anniversaries, many more features were named for her. And long after her death, Queen Peak in northern British Columbia was named for her in 1933 because of its associations with nearby Victoria Peak and Consort Peak.

The best-known place named for the British monarch is the beautiful city at the southern tip of Vancouver Island. In 1843, the Hudson's Bay Company resolved to name the new fort overlooking Juan de Fuca Strait for her. Briefly, Fort Albert was substituted, but a terse message from London insisted on the use of Fort Victoria. The townsite of Victoria was established at the fort in 1851–2, and in 1868 the growing city became the capital of the colony of British Columbia, which was to become a province three years later.

Alberta had a Victoria northeast of Edmonton where George McDougall established a mission in 1862 and the Hudson's Bay Company set up a post two years later. In 1887, to avoid confusion with other Victorias, this small community was renamed Pakan, the nickname of a Cree chief. The village of Empress, northeast of Medicine Hat, was named in 1913 in commemoration of the Queen's imperial title received from Parliament in 1876 when Benjamin Disraeli was prime minister.

The Marquess of Lorne and his wife, Princess Louise (the Queen's daughter), originally wanted to give the name Victoria to the capital of the North-West Territories in 1882, but wisely chose the other half of her Latin title, Regina.

Manitoba has both a rural municipality and a lake named Victoria, as well as another municipality called Victoria Beach.

In terms of geographical names, Ontario is Canada's most 'Victorian' province. It has at least forty-seven distinct features with her name: one county, one township, fourteen populated places, and thirty-one physical features. In fact, one does not travel far in Ontario before encountering Victoria Corners, Victoria Square, Victoria Harbour, Victoria Springs, Victoria Lake, or just plain Victoria.

Evidence of Victoria is less apparent in Québec, although the second-largest place in Canada with her name is in that province. Victoriaville, a city of 38,200 people, was named for the Queen in 1861. There are as well seven physical features in Québec with the name Victoria, including Grand lac Victoria at the head of the Ottawa River south of Val d'Or.

The Atlantic provinces have twenty-nine places and features with the name Victoria. Among these are a county in each of New Brunswick and Nova Scotia. Victoria is an attractive seaside village in Prince Edward Island, where there are also places called Victoria Cross and Victoria West. Newfoundland has a Victoria, a town of 1,381; it lies on the west side of Conception Bay about 50 kilometres northwest of St. John's.

Our three northern territories have twenty-two features with the name Victoria. Among them are Victoria Island, Canada's second-largest island (after Baffin) in the Arctic Archipelago, and Victoria and Albert Mountains on Ellesmere Island.

Canadian Tributes for Four Queen Elizabeths

Four Queen Elizabeths have been honoured in Canada's toponymy. Monarchists know there have been only three Elizabeths who have been English sovereigns or consorts since the beginning of British exploration on our Atlantic coast in the early years of the sixteenth century. There is, however, a Mount Queen Elizabeth named after a fourth queen – but more about her later.

Queen Elizabeth Foreland was the first European

name given in the Canadian Arctic. Explorer Martin Frobisher set out in the summer of 1576 to sail through the perceived Northwest Passage to Cathay, and after passing Greenland, sighted a 'highe lande' on 20 July, and named it 'after hyr Majesties name ...'

Frobisher then entered a body of water whose shores to the right he judged belonged to Asia, and those to the left he assumed was the 'firme' of America, and gave it the name of 'Frobisher's Streytes.' After his third trip in 1578, he concluded he had only entered a dead-end bay, now Frobisher Bay, far from the shores of Cathay.

The precise location of the foreland was not clearly established in the published reports of Frobisher's travels. In 1910, the Geographic Board decided it should be placed on the southeast side of Loks Land, an island on the *north* side of Frobisher Bay, just offshore from Baffin Island. This decision was confirmed in 1957.

Noted American author and admiral Samuel Eliot Morison, in *The European Discovery of America, The Northern Voyages, AD 500–1600*, published in 1971, suggested that the geographical and navigational evidence revealed that the foreland was the southeast cape of Resolution Island, on the *south* side of Frobisher Bay and on the north side of what Frobisher called 'Mistaken Straites,' called a century later Hudson Strait for another famous explorer. Although Adm. Morison did not question the Canadian decision (if he even knew about it), his conclusion as to the actual location of the foreland seems more realistic. But after a hundred years on Canadian maps, it would be surprising to find the name relocated to Resolution Island.

Queen Elizabeth, who herself never travelled outside England, was so taken with Frobisher's descriptions of the new-found lands, she christened them Meta Incognita, 'limits unknown.' In 1957, the Canadian Board on

Geographical Names gave the name Meta Incognita Peninsula to the 320-kilometre-long body of land separating Hudson Strait from Frobisher Bay.

In 1580, Queen Elizabeth issued the fundamental principle of the freedom of the seas, declaring that 'neither can any title to the Ocean belong to any people or private man, forasmuch as neither nature, nor regard of the public use permitteth any possession thereof.' In commemoration of this principle, British Columbia surveyor-general George Aitken proposed in 1933 the naming of Mount Queen Bess for her. The mountain, with a elevation of 3,290 metres, is north of the head of Bute Inlet, 245 kilometres northwest of Vancouver.

In the spring of 1939, King George VI and the highly popular and gracious Queen Elizabeth (the Queen Mother after 1952), undertook a very strenuous tour of Canada and the United States by train, just a few months before the outbreak of the Second World War. On 29 May, just eleven days after arriving at Québec City, they were in Vancouver. A 52-hectare park in the centre of the city, which had been set aside as a reserve in 1912, was dedicated in honour of the queen. Queen Elizabeth Park, with an arboretum, gardens, and sports fields, occupies the highest point in central Vancouver.

The Royal couple returned to Central Canada, where, on 7 June, as 100,000 turned out in the city of St. Catharines to catch a glimpse of them, the Royal automobile broke an electric beam on Canada's first four-lane highway, and officially inaugurated the Queen Elizabeth Way. Variously called The QEW and The Queen E, the controlled-access highway was built between Toronto and Burlington along the Middle Road between Highway 2 and Dundas Street, across the land barrier called Burlington Beach, and then along the Lake Ontario shoreline to St. Catharines, where it climbed the Niagara

Escarpment to Niagara Falls. It was completed to the Peace Bridge in Fort Erie in 1956.

The most popular among the Queen Elizabeths in terms of Canadian geographical features is the highly esteemed Queen Elizabeth II. A month prior to her coronation on 2 June 1953, the Geographic Board of Alberta proposed giving the name Queen Elizabeth Ranges to the towering mountains surrounding Jasper National Park's picturesque Maligne Lake. The subsequent announcement by the Canadian Board on Geographical Names drew attention to the perfect picture of alpine grandeur, with 'bold rocky forms, ice and snow gleaming against a blue sky, dark forests and a sapphire blue lake – a fitting memorial to the Queen.'

Patrick Baird, leader of the Arctic Institute's 1953 expedition, proposed the name Mount Queen Elizabeth for a spectacular peak overlooking the Inuit village of Pangnirtung, on the west side of Baffin Island's Cumberland Peninsula. His proposals for Coronation Fiord and Coronation Glacier were accepted for features on the east side of the peninsula, but Governor General Vincent Massey judged the mountain too insignificant as an honour for the new monarch.

Geologist Hugh Bostock suggested Elizabeth Mountains for the massive alpine region of Ellesmere and Axel Heiberg islands, the most mountainous region east of the Canadian Cordillera. Before the merits of Dr. Bostock's suggestion could be reviewed, Graham Rowley, then director of northern administration and lands in the Department of Northern Affairs and Natural Resources, proposed Queen Elizabeth Group for all the islands north of the deep bathymetric trench from Lancaster Sound in the east through Viscount Melville Sound to M'Clure Strait on the west. The names board amended the name to Queen Elizabeth Islands. The department's deputy minister, Gordon Robertson, obtained Vincent Massey's endorsement, and subsequently its minister, Jean Lesage, announced the naming of the islands in the House, stating that 'in the reign of her majesty Queen Elizabeth II, we have every confidence we are on the threshold of a new age, if not of discovery, then of development of our Arctic lands and seas.' Geographer Andrew Taylor described the islands in the *Canadian Geographical Journal* (June 1956) as 'isolated, naked and rugged, set in a sea of ice, they form the northern bastion of the continent.'

One of the most attractive boulevards of the nation's capital is the route from Parliament Hill to Dows Lake along the west side of the historic Rideau Canal. Called simply The Driveway when it was opened in 1904, it was christened Queen Elizabeth Drive in 1967 during a visit by the queen to Canada. However, The Driveway continues to be the preferred designation among local residents.

Montréal's largest hotel, with 1,143 rooms and suites, is the stately Queen Elizabeth Hotel, opened in the spring of 1958 by the Canadian National Railways, but now operated by Canadian Pacific. Its designation in French, Le Reine Elizabeth, often draws puzzled looks from francophiles, until they are told the article represents the masculine noun 'hotel.'

Queen Elizabeth II has been honoured in the names of several schools and hospitals and other public facilities across Canada. In 1959, a civic auditorium was built in downtown Vancouver, and named Queen Elizabeth Theatre in honour of the queen. Canada's first planetarium was the Queen Elizabeth Planetarium, opened in 1962 in Edmonton. In 1978, Lac Cardinal Provincial Park near Grimshaw in Alberta's Peace River country was renamed Queen Elizabeth Park to commemorate a visit by the queen.

The queen's name is also recalled in the Queen Elizabeth II Coronation Medal and the Queen Elizabeth II Silver Jubilee Medal, struck in 1953 and 1977 as Canadian awards for outstanding public service.

Now for the fourth queen. At the very south end of Banff National Park, and on the Alberta–British Columbia boundary, at a height of 2,850 metres, is Mount Queen Elizabeth, and beside it is Mount King Albert. These names, and those of several other figures prominently associated with Allied victories during the First World War, were proposed in 1917 by interprovincial boundary commissioner Arthur O. Wheeler. Elisabeth (1876–1965), a daughter of the Duke of Bavaria, became the wife of Belgian Prince Albert in 1900. Albert became king in 1909, and he and Elisabeth reigned as king and queen of the Belgians until 1934, when Albert died after a fall while rock climbing.

Fit for the Prince of Wales

A long-established practice in Canada since the early 1700s has been the naming of physical features, museums, theatres, and commercial establishments for the current Prince of Wales. The name has also been given to a National Hockey League conference (from 1974–5 to 1992–3), as well as a prestigious annual horse race in Fort Erie, Ont.

The first Prince of Wales to be commemorated in Canada was the son of King George I. Born in 1683, he later ruled Britain for thirty-three years as King George II. The name Prince of Wales's Fort was given, in 1719, to the Hudson's Bay Company's stone-walled post at the mouth of the Churchill River in northern Manitoba. In 1782, a French fleet under the command of Comte de

Lapérouse blew up Fort Prince of Wales, as it came to be known. In the 1930s, the fort was restored. Now a national historic site, it is one of the main attractions in the Churchill area.

Perhaps the most distinguished visitor to North America in 1860 was Prince Albert Edward, the eldest son of Queen Victoria and great-great-grandfather of the present Prince of Wales, Prince Charles. He received a tumultuous welcome from the time he arrived in St. John's, Nfld., on 24 July until his departure from Portland, Me., on 20 October. Even before his arrival, the Prince of Wales had been recognized in at least five physical features named for him. Prince of Wales Island, located 1,320 kilometres north of Yellowknife, is perhaps the best known, but it was not the first island in the Arctic to bear his name. When explorer and trader John Rae crossed Melville Peninsula to the Gulf of Boothia in 1846–7, he named an island in the gulf's Committee Bay for the then five-year-old prince. In 1926, the Geographic Board of Canada renamed it Wales Island to avoid confusion with the larger Prince of Wales Island to the west.

In 1850–1, Robert M'Clure wintered in a 275-kilometre-long passage between Victoria Island and Banks Island. He named it Prince of Wales Strait. Although it is now considered part of the main channel of the Northwest Passage, its first recorded navigation was not until Henry Larsen's RCMP patrol in 1944. In 1852, a high range on Ellesmere Island was named Prince of Wales Mountains after Prince Albert Edward. Finally, Prince of Wales Reach was assigned in 1859 to a long, narrow passage of Jervis Inlet, 90 kilometres northwest of Vancouver.

After visits to Halifax and Saint John in 1860, the prince and his retinue set out for Fredericton. From Saint John, they travelled 14 kilometres by train before boarding a steamer on the Kennebecasis River at Rothesay, newly named for one of the prince's titles, the Duke of Rothesay.

The community of Prince of Wales, 18 kilometres southwest of Saint John, may have been named about this time. Its choice was perhaps reinforced by the settlement of that area in the late 1780s by the Prince of Wales American Regiment, Loyalist soldiers who were granted land there after the American Revolution. The Prince of Wales post office operated there from 1861 to 1950.

After visiting Charlottetown, where a college was named in his honour, the prince went up the St. Lawrence River to Québec City and Montréal. On 29 August, he travelled on the Grand Trunk Railway to Dickinson's Landing, 14 kilometres west of Cornwall, Ont., then returned to Montréal that evening after shooting the Long Sault Rapids. That year, the Grand Trunk named its Dickinson's Landing station Wales.

In 1860, a new road leading north from Orangeville, Ont., was being cut through the woods. On completion of the road, one of the workmen is said to have exclaimed that it was fit for the Prince of Wales – to this day the name remains in use. Streets were also named for the Prince of Wales in St. John's and Montréal.

The second Prince of Wales to make a popular tour of Canada was the eldest son of George V – the son who became Edward VIII, the Duke of Windsor. While the prince was visiting Ottawa in 1919, he endorsed a proposal to name the falls at Hogs Back on the Rideau River for him. At the same time, the provincial highway from Ottawa to the St. Lawrence River at Johnstown, Ont., was officially named The Prince of Wales Highway. Neither name gained public recognition or acceptance. Prince of Wales Falls remains the official designation on maps and in publications about the Rideau River and the Rideau Canal, but it is doubtful anyone outside official naming and mapping offices is aware of it. Although the name of the highway never caught on, that part of Highway 16 between Carling Avenue and the former city limits just south of Hog's Back Road acquired the name Prince of Wales Drive – a name well known in the national capital region. The Prince of Wales Hotel, the largest all-wood structure in Western Canada, was opened in 1927 in Waterton Lakes National Park.

The current Prince of Wales has been honoured by the Prince of Wales Northern Heritage Centre, a museum of Aboriginal artifacts in Yellowknife which opened in 1979.

Vanier: Places Honouring an Illustrious Statesman

When the Royal Canadian Geographical Society moved its headquarters from Wilbrod Street in Ottawa to McArthur Avenue in Vanier in 1988, some Society members wondered how far the Society had moved from the nation's capital. Most were surprised to learn it was only about half a kilometre to the east.

Located on the eastern bank of the Rideau River, 2 kilometres from its mouth, Vanier was originally settled in the 1830s, when developer Charles Cummings built a bridge across the river. The growing community on Montreal Road was called Cummings Bridge for almost half a century. In the 1870s, it was renamed Janeville, either in honour of the wife of another settler of the 1830s, Donald McArthur (hence McArthur Avenue), or for the first child born in the community.

The name Eastview was given on 31 December 1908, when the communities of Janeville, Clarkstown, and Clandeboye – in what was then Gloucester Township – were incorporated into a single village with an area of

2.79 square kilometres. Eastview – the name was chosen because there is a striking view of the Parliament Buildings from its eastern limits – became a town in 1912 and a city in 1963. In 1950, Eastview applied to the Ontario Municipal Board to annex adjacent lands in Gloucester Township, but the board gave approval to Ottawa instead. That left Eastview surrounded by Ottawa and set the stage for Vanier's distinction as Canada's smallest city, which it lost on 1 January 2001, when it was amalgamated with the city of Ottawa.

On 1 January 1969, Eastview was renamed Vanier, primarily to honour the first French Canadian to be appointed governor general, but also to give the city a name with francophone character. Fifty-five per cent of its population of 18,150 is francophone, its street names are largely French, it is the headquarters of the Association canadienne-française de 1'Ontario and other francophone groups, and its distinctly French-Canadian atmosphere is widely known throughout the national capital area.

Vanier's population had steadily declined over the past thirty years as residents left behind their small, quaint homes on narrow streets for larger houses and lots in the suburbs. Down from 24,700 in 1963, Vanier had the smallest population of the five cities in the former Regional Municipality of Ottawa–Carleton (the others were Gloucester, Kanata, Ottawa, and Nepean), before the region became the city of Ottawa on 1 January 2001.

More than once provincially appointed commissions had urged amalgamation with Ottawa. Even in 1984, a majority of Vanier's voters recommended joining the capital city, though Vanier city council at the time declined to act. Ottawa had provided Vanier with police service and fire protection but, worried about increased costs, resisted annexing its small neighbour, whose median income was only 70 per cent of Ottawa's. More recently, Vanier had displayed a renewed pride in itself, and was determined in the 1990s to go its own way.

Georges Philias Vanier was one of the most admired governors general to hold that office during Canada's first century. Born in Montréal in 1888, he studied law at Université Laval before becoming a career soldier. He was one of the founding officers of the 22nd Battalion of the Canadian Expeditionary Force, and lost a leg during a First World War battle in France. After the war, the battalion became the Royal 22nd Regiment (the illustrious Vandoos, an anglicization of vingt-deux), and Vanier served as its commanding officer from 1926 to 1928. He distinguished himself in Canadian diplomatic postings in London, Geneva, and Paris, and recruited French Canadians to serve in the Second World War. In 1959, he was appointed governor general, the Queen's representative in Canada. In the third month of Canada's centennial celebrations, the man who had become renowned for his love of Canada, and for his deeply spiritual and moral dimensions, died at Rideau Hall.

Ontario's Vanier has not been the only city in Canada named for the governor general. In 1966, the Québec City suburb of Québec-Ouest, with an area of 4.68 square kilometres and a population of about 11,000, was renamed Vanier. When the legislation to change Eastview's name was being considered in Ontario in 1968, officials in Vanier, Qué., argued that only one Canadian city should have the name, but the wishes of the Ontario city's residents prevailed.

In Québec, four lakes, a township, a creek, a hill, and an electoral division are named for Vanier. Forty towns and cities in the province have a Vanier street, boulevard, or avenue. Elsewhere in Canada, there are another twenty-six towns and cities with a street, avenue, or drive named for the former governor general.

In the 1950s, what was thought to be a peninsula of Bathurst Island in the Arctic Archipelago was found to be three large islands and two smaller ones. Initially, the names of three American vessels involved in establishing early-warning radar stations across the High Arctic were given by geographer Andrew Taylor to the large islands, but they were rejected in 1960 by the Canadian Board on Geographical Names. The board proposed naming the islands after Vanier and his two predecessors, Vincent Massey and Viscount Alexander of Tunis. In 1962, the federal cabinet approved the proposals and stated the choices would reaffirm Canadian sovereignty in the Arctic. The form Île Vanier was adopted for the largest of the three islands, at 970 square kilometres.

A year later, the small island between Île Vanier and Massey Island was named Île Pauline for the popular and gracious wife of the governor general. Vanier replied that his wife was pleased indeed that the 'baby' island was next to his. Knowing that there was a larger unnamed island between Alexander and Massey islands, he quipped that after 'forty-two years of marriage she still prefers my company.'

As Robert Speaight noted in his book, *Vanier: Soldier, Diplomat and Governor General* (1970), Vanier also mused whimsically about the appointment of an island manager for Île Vanier, the design of an island flag, and the publication of two newspapers, one in English and one in French.

In 1967, the Whitehorse Lions Club proposed naming a mountain overlooking the north end of Kusawa Lake, 55 kilometres west of Whitehorse, for the governor general. A four-member team climbed the mountain on 9 July and erected a cairn at its summit. Some badges from General Vanier's Royal 22nd uniform and some photos of the late governor general were embedded in the cairn by the regiment's Staff Sgt. Albert Bélanger. The Canadian Permanent Committee on Geographical Names endorsed the names three days after the climb. Mount Vanier has an elevation of 1,845 metres and is visible from the Alaska Highway, 20 kilometres to the north.

Also in 1967, a Royal Air Force expedition from Britain climbed a mountain near the north end of Ellesmere Island and named it for Vanier. This second Mount Vanier in Canada's north, approved in 1971, has an elevation of 2,417 metres, just 200 metres lower than nearby Barbeau Peak, the highest mountain in the Canadian Arctic.

The name Vanier is recalled in many other ways in Canada. The Vanier Institute of the Family was established in 1964 by Georges and Pauline Vanier to promote research into the family and study the complexity of modern living and its effects on the integrity of family life. Since 1960, the Institute of Public Administration of Canada has annually awarded the gold Vanier Medal to an individual showing outstanding leadership in Canadian public administration.

A community college in Saint-Laurent, Qué., a western suburb of Montréal, is named Vanier College. Vanier Park in Vancouver is the site of the City of Vancouver Archives, and the Centennial Museum and Planetarium. Nationwide, there are many elementary and secondary schools bearing his name. And each November since 1965, the best of Canada's university football teams compete for the Vanier Cup.

Reserving Lofty Peaks for Our Prime Ministers

The year 1927 marked a high point in Canada's development as an independent state among the world's sovereign nations. Canada had reached its diamond jubilee of Confederation – sixty years of growing independence from the government of Great Britain. It was just the year before that Canada had been allowed to designate its own ambassadors to foreign countries, and to have its governor general appointed by the Crown and not by the British government. More independence was still to be achieved during the remaining years of the twentieth century – and some will say we still have too many restraining ties to Britain – but 1927 was a year to celebrate the birth of a country.

Parties and celebrations were organized from coast to coast. Parades with marching bands and decorated floats were routed through cities, towns, and villages. Speeches trumpeted the glorious flowering of this great land. Ideas were solicited to recognize achievements attained.

Suggestions to name either a prominent geographical feature for Prime Minister Mackenzie King, or impressive features for all of Canada's ten prime ministers who had served from 1867, were first proposed in May 1927 by the Modern History Club of Victoria High School. George G. Aitken, the British Columbia member of the Geographic Board of Canada, liked the second idea, but pointed out that of the ten prime ministers only four – Sir John Abbott, Sir John Thompson, Sir Mackenzie Bowell, and Mr. King – had not already been honoured with the naming of a significant feature in the province. Mountains rising above Rogers Pass had been named in 1887 for Sir John A. Macdonald, Alexander Mackenzie, and Sir Charles Tupper. Mount Laurier in the Yukon had been named in 1890, and another Mount Laurier, proposed in 1913 in north-central British Columbia, was accepted in 1928. Mount Sir Robert had been given in 1916 in the Coast Mountains for Sir Robert Borden. But Aitken should have said five out of the ten, because no major feature had been named for Arthur Meighen.

In July 1927, the Geographic Board endorsed the idea of naming peaks for the four prime ministers listed by Aitken, and added the names of Sir Wilfrid Laurier and Arthur Meighen. (By adding Laurier's name it appears the board was currying favour with King, Laurier's successor as Liberal leader.) The board reserved an area with fifteen peaks exceeding 10,000 feet (3,050 metres) in the southern part of the Cariboo Mountains, on the west side of the Rocky Mountain Trench, some 90 kilometres west of the Yellowhead Pass. It was named Premier Group, 'premier' being a common word used at the time for the prime minister of each of the British dominions.

The highest peak at 11,750 feet (3,580 metres) had been named Mount Titan in 1924 by American alpinists Allen Carpe and Rollin Chamberlin, but the Geographic Board substituted Mount Sir Wilfrid Laurier for it. Another at 11,250 feet (3,430 metres) near the head of the North Thompson River had been named by Carpe for explorer David Thompson, but it was renamed Mount Sir John Thompson. Kiwa Peak, also at 11,250 feet, at the head of Kiwa Creek, was changed to Mount Sir John Abbott. Mount Sir Mackenzie Bowell, at 11,000 feet (3,353 metres), had been named Mount Welcome by Carpe.

The Geographic Board hesitated in 1927 to choose peaks for Arthur Meighen and Mackenzie King, as both were still living. It concluded it would be too difficult to select peaks of relative elevation for the two political adversaries. It was not until 1962, two years after the death of Meighen and twelve after King's death, that the choosing of peaks for the two prime ministers came up again. With a height of 10,600 feet (3,230 metres), a reserved peak was named Mount Mackenzie King by the newly created Canadian Permanent Committee on Geographical

Names. In 1925, climber Don Munday had named it Mount Hostility. Another peak, at 10,100 feet (3,078 metres), which an American climbing team had named after Allen Carpe in 1949, was given the name Mount Arthur Meighen. Mount Richard Bennett, with a height of 10,400 feet (3,170 metres), was named at the same time for the Conservative prime minister who governed for five years during the Great Depression. The Permanent Committee also renamed the group Premier Range.

In August 1927, British Columbia's acting premier, Dr. John D. MacLean, requested the naming of a mountain in the Premier Group for British prime minister Stanley Baldwin, who was then on a tour of Canada. The feature named Mount Challenger by Carpe was chosen by the Geographic Board for Mount Stanley Baldwin. Its height is 10,660 feet (3,249 metres).

In late 1927, Dr. MacLean proposed naming a mountain in the group for John Oliver, premier of British Columbia from 1918 to his death in August 1927. A mountain, with the height of 10,500 feet (3,200 metres) and visible from the Mount Robson and Tête Jaune Canadian National sidings, was selected by the province, and approved by the Geographic Board. Called Mount Aspiration in 1925 by climber Don Munday, Mount John Oliver is the farthest north of the peaks in the range named for a first minister of a government, and the only one named for a British Columbia premier.

The renaming of these mountains did not sit well with the mountain-climbing community. The most distinguished among them in the 1920s, American alpinist Dr. J. Monroe (Roy) Thorington, claimed the Geographic Board was violating its own rules of priority publication and local usage. The board's secretary, Robert Douglas, acknowledged the valuable contributions of Carpe and Chamberlin, but pointed out that they had not submitted their names to the board for endorsement, so that the features were considered officially unnamed. Other alpinists regretted leaving significant features unnamed, while lesser features lower than 10,000 feet in the Premier Range (e.g., Mount Milton, Penny Mountain) were being named.

In 1963, Arthur Laing, then minister of Northern Affairs and Natural Resources, suggested naming a peak for former prime minister Louis St-Laurent, who had retired from public life in 1958. Prime Minister Pearson endorsed the proposal. A peak said to be 10,365 feet in height, but later found to be only 9,900 feet (3,017 metres), was selected, and officially named Mount Louis St-Laurent in October 1964.

Lester Pearson himself died in December 1972. An unnamed peak rising to 10,500 feet (3,200 metres) was chosen to honour the distinguished diplomat and head of the government of Canada. Mount Lester Pearson was proclaimed in September 1973.

The following year, the name of a premier from another era was put forward by Barbara McNabb of Edmonton. She suggested a peak be named for Sir Allan Napier MacNab, who had been co-premier of Canada from September 1854 to May 1856. Don F. Pearson, the British Columbia representative on the Permanent Committee, gave the name Mount Sir Allan MacNab to the most southeasterly mountain of the range, which rises only to 7,750 feet (2,360 metres).

On the death of John Diefenbaker in August 1979, American mountain climber Richard Estock, who had climbed in the Premier Range earlier that summer, reported that other climbers were proposing to rename a mountain, unofficially called Pyramid Peak in the alpine literature, for him. Because a huge reservoir on the South Saskatchewan River had been named Lake Diefenbaker in 1967 – but more likely because of strained relations between the British Columbia people and the federal Liberal government – Estock was informed by Don Pearson that there were no plans at that time to name a peak in the Premier Range for Diefenbaker, and, by implication, for any of his successors.

Five peaks with elevations exceeding 10,000 feet (3,050 metres) remain unnamed in the Premier Range. Decisions to name any them for Diefenbaker and any of his successors – and even for provincial premiers – rest entirely with the names authority of British Columbia.

Administrations of Canada's Prime Ministers

Sir John A. Macdonald (1815–1891)	1867–73; 1878–91
Alexander Mackenzie (1822–1892)	1873–8
Sir John J.C. Abbott (1821–1893)	1891–2
Sir John S. Thompson (1845–1894)	1892–4
Sir Mackenzie Bowell (1823–1917)	1894–6
Sir Charles Tupper (1821–1915)	1896
Sir Wilfrid Laurier (1841–1919)	1896–1911
Sir Robert L. Borden (1854–1937)	1911–20
Arthur Meighen (1874–1960)	1920–1; 1926
W.L. Mackenzie King (1874–1950)	1921–6; 1926–30; 1935–48
Richard B. Bennett (1870–1947)	1930–5
Louis S. St-Laurent (1882–1973)	1948–57
John G. Diefenbaker (1895–1979)	1957–63
Lester B. Pearson (1897–1972)	1963–8
Pierre E. Trudeau (1919–2000)	1968–79; 1980–4
C. Joseph Clark (1939–)	1979–80
John N. Turner (1929–)	1984
M. Brian Mulroney (1939–)	1984–94
Kim Campbell (1947–)	1994
Jean Chrétien (1934–)	1994–

A Mountain for Michener, a Lake for Léger

During the first 100 years of Confederation, more than 100 geographic features – and countless more parks, schools, bridges, and institutions – were named for the nineteen governors general of Canada. Among those most honoured in geographic names are the Earl of Dufferin (eighteen) and the Marquess of Lorne (fourteen). But the story behind the naming of features for the twentieth and twenty-first governors general, Roland Michener and Jules Léger, is one of the more unusual ones.

Daniel Roland Michener, governor general from 1967 to 1974, was born 19 April 1900, in Lacombe, Alta., 130 kilometres south of Edmonton, and raised in Red Deer, 25 kilometres farther south. He had a distinguished career as a parliamentarian, Speaker of the House of Commons, and High Commissioner to India.

Until he died in 1991, Michener was an active senior partner in the Toronto legal firm Lang, Michener, Lawrence and Shaw, and he still followed the vigorous exercise routine that made him one of the more memorable advocates of Participaction, the organization that promotes fitness to Canadians.

In 1977, a proposal was sent by the secretariat of the Canadian Permanent Committee on Geographical Names to the Alberta Historic Sites Board (the body then responsible for geographic naming in the province) to name a feature for Michener. The board greeted the suggestion with enthusiasm and proposed it for a prominent mountain 185 kilometres west of Red Deer.

The mountain is located on the south side of Abraham Lake along the upper reaches of the North Saskatchewan River. Its magnificent profile was seldom seen by travellers until the early 1970s, when the David Thompson Highway was built, linking the town of Rocky Mountain

House to the Icefields Parkway in Banff National Park. The mountain has an elevation of 2,545 metres and rises 1,220 metres above the lake.

Before proceeding with the naming, the board consulted the Stoney tribal council, as the surrounding territory is traditional Stoney land. Although the Stoney know the feature as *îyârhe kta ûtha*, meaning 'person who dwells in the mountains,' they endorsed the idea of Mount Michener as its official name.

The naming ceremony took place at Windy Point, on the north side of the lake, on 26 August 1979. After unveiling a plaque, then-premier Peter Lougheed observed, 'We think this mountain ... will be a continual reminder to Albertans and visitors to our province of the outstanding contributions made by Mr. Michener to the life of this country.'

On learning that Michener had been seen jogging beside the highway before the ceremony, Lougheed remarked that he almost expected the athletic 79-year-old to cross the lake and climb to the summit of the mountain. Perhaps that suggestion inspired Michener, because nearly three years later he did climb to the top, accompanied by two mountain guides.

Author Peter Stursberg, in *The Last Viceroy* (1989), claimed that Michener set 'a record as the only person after whom a mountain was named who ever reached its peak.' If he had substituted 'octogenarian' for 'person,' his claim would likely be unassailable.

In 1976, Michener was among the guests of his successor, Jules Léger, during one of Queen Elizabeth's visits to Canada. During an informal conversation, Léger asked Michener if he knew of a tradition of naming features for governors general during their terms of office, and he understood Michener to say that, indeed, there had been a feature named for him in the North. (In fact, what had been named, earlier in 1976, was the Norah Willis Michener Game Preserve on the Northwest Territories–Yukon boundary – for Michener's accomplished wife.)

Acting on the knowledge that an Arctic island had been named in 1962 for Georges Vanier, and assuming a feature had been named for Michener while he was in office, Government House contacted the Canadian Permanent Committee on Geographical Names to find out when a feature would be named for Léger.

The response was that there was no such established custom. In fact, the committee and most provincial names boards had taken a fairly rigid stance in the 1970s against naming features for living persons. Mount Michener was among the last to be allowed.

When Léger learned in 1979 that a mountain had been named for Michener, he asked when a feature would be named for him, adding that he would prefer a lake in his native Québec. His suggestion was sent to the Commission de toponymie du Québec, but because of the commission's rigid adherence to its principle against naming features for living persons, a decision was deferred.

Léger was born in 1913 in Saint-Anicet, on the south shore of Lake St. Francis, a widening of the St. Lawrence River 60 kilometres southwest of Montréal. After a brilliant diplomatic career, he served as under-secretary of state before being appointed governor general in 1974. Although a stroke impaired his speech and mobility, Léger continued to perform the duties of his office with the quiet dignity for which he had become widely known over the years.

Léger died in Ottawa in November 1980. Two years later, the commission gave the name Lac Jules-Léger to a lake in north-central Québec near the great divide separating the headwaters of the St. Lawrence from the Hudson Bay drainage basin. At her home in Ottawa,

Gabrielle Legér, the governor general's widow, was presented with a laminated map of the region surrounding Lac Jules-Léger.

Laurier: Most-Commemorated Canadian

The individual most honoured by names on our maps is Queen Victoria. However, the most-commemorated Canadian is Sir Wilfrid Laurier, who has thirty-seven geographical features named for him, to say nothing of dozens of schools, parks, and streets from Newfoundland to the Yukon.

Laurier – born in 1841 in Saint-Lin, Qué., 35 kilometres northwest of Montréal – was prime minister of Canada from 1896 to 1911. During this time, the Geographic Board was set up to coordinate naming in Canada. Before he died in 1919, at a time when naming for living persons was common, at least eleven communities and features had been named for him. By 1928, the total was up to twenty, and seventeen more were added in subsequent years.

Widely respected in both English and French Canada, Laurier has been commemorated in the names of twelve features in Québec, seven in Ontario, six in British Columbia, seven in the Prairie provinces, three in the Territories, and two in the Atlantic provinces.

In 1896, geologist Robert Bell explored a vast area of west-central Québec surrounding Lac Matagami. South of the lake he observed a mountain, the most conspicuous feature of the landscape, and named it Mount Laurier (officially it is now Mont Laurier) for the recently elected prime minister.

From 1859 to 1902, the area of Laurierville, 65 kilometres southwest of Québec City, was part of the civil parish of Sainte-Julie-de-Somerset. In 1902, the village was incorporated and named for the prime minister. It now has a population of 885.

When the Intercolonial Railway was built from Saint-Hyacinthe to Lévis, one of the stations on the line was Saint-Flavien. Located 48 kilometres southwest of Lévis, the station was renamed Laurier when the Intercolonial became part of she Canadian National Railway system. A village that developed around the station was incorporated as Laurier-Station in 1951. Situated beside the busy Trans-Canada Highway, it now has a population of 2,175.

The area of the present town of Mont-Laurier, 175 kilometres northwest of Montréal, was settled in the late 1800s. The community of Rapide-à-l'Orignal had been established at a stretch of rough waters on the Rivière du Lièvre. In 1909, Curé Génier, the parish priest, urged Québec's premier Lomer Gouin to designate the village as the centre of the new judicial district being planned in the area, and enlisted the support of the prime minister in his cause. As a result, the village was incorporated as Mont-Laurier. Elevated to town status in 1950, Mont–Laurier is now an active commercial and tourist centre, with a population of 7,865.

In 1934, the name Mont Laurier was proposed for a mountain, 16 kilometres north of the community. Pointing out that another mountain near Lac Matagami was already named that, the authorities substituted Mont Sir-Wilfrid.

About 40 kilometres east of La Tuque (north of Trois-Rivières) is a township named Laurier. Within the township are Lac Laurier and Lac Wilfrid, both named in 1948. There is another Lac Laurier in Parc des Laurentides, 35 kilometres south of Chicoutimi. In Parc de la Vérendrye northwest of the town of Mont-Laurier, there is another Lac Laurier, so named in 1973.

In New Brunswick, a post office (now closed), 35 kilometres northeast of Chatham, was named Saint-Wilfred in 1932 in honour of the former prime minister, although his name was spelled incorrectly. In 1958, a small islet in Lockeport Harbour, N.S., was named Laurier Rock.

Laurier Township in the Parry Sound District of Ontario was named in 1878 when Laurier was minister of inland revenue. This appears to be the earliest use of his name for a feature. Within the township, Fox Lake was renamed Laurier Lake in 1951.

At the point where the Spanish River flows into Lake Huron, there are two small islands called Laurier Island and Wilfrid Island. About 20 kilometres northeast of Kenora is a small lake called Laurier Lake, which may have been named at the turn of the century during a railway survey. The lake is drained by Laurier Creek.

Among settlers in an area 45 kilometres southeast of Dauphin, Man., were some Québec farmers who were admirers of Wilfrid Laurier. When Laurier was first elected prime minister in 1896, the name Laurier was substituted for Fosbery.

Laurier Island in Cree Lake in northern Saskatchewan was named in 1938. Laurier Rural Municipality in the area of Radville, Sask., 110 kilometres southeast of Regina, was also named for Sir Wilfrid in 1909.

Alberta also has a Laurier Rural Municipality named for Sir Wilfrid. The municipality, west of the town of St. Paul and 130 kilometres northeast of Edmonton, was named in 1914. There is a Laurier Lake 60 kilometres east of St. Paul.

The earliest use of Sir Wilfrid's name in British Columbia was in 1897, when J.D. Moodie of the North West Mounted Police took an inspection trip from Edmonton to the Yukon. He named Laurier Pass, which traverses the Rocky Mountains 190 kilometres northwest of Fort

St. John. In 1913, the peak on the south side of the pass was named Mount Laurier by F.K. Vreeland of New York. Although the Geographic Board was reluctant to have another mountain named for the prime minister, the name was adopted fifteen years later. At the same time, Mount Lady Laurier, 25 kilometres west of the pass, was named for his wife.

A year earlier, in 1927, a mountain rising to 3,580 metres in the Premier Range in British Columbia was named Mount Sir Wilfrid Laurier. The range is directly west of Valemount and 50 kilometres south of Mount Robson, at 3,942 metres the highest peak in the Rockies. Several Canadian prime ministers, and some other first ministers of British Columbia and Great Britain, have been honoured in the names of mountains in the Premier Range rising above 3,050 metres (see also pp. 299–304).

Laurier Cove in Tuck Inlet north of Prince Rupert may have been named in 1906 during a survey of the inlet, or in 1910 when the prime minister visited the western terminus of the Grand Trunk Pacific.

In 1888, surveyor William Ogilvie named a mountain overlooking Lake Laberge, 25 kilometres north of Whitehorse, for the indefatigable scientist and explorer George M. Dawson. Two years later, Dr. Dawson insisted that the name be changed from Mount Dawson to Mount Laurier. Later, in 1936, Laurier Creek, which flows west into Lake Laberge, was also named for Sir Wilfrid.

In Québec, two-thirds of the current 89 postal divisions have a thoroughfare named for Sir Wilfrid. Elsewhere in Canada, one-quarter of the 201 postal divisions have a street commemorating the former prime minister. Prominent among them is Laurier Avenue in Ottawa, where Sir Wilfrid owned a large house, now a museum.

Among schools, the most noteworthy is Wilfrid Laurier University in Waterloo, Ont. Founded as Waterloo

College, and named Waterloo Lutheran University in 1957, it was renamed for the former prime minister in 1973 when it became a provincially funded institution. There are secondary schools in Ottawa, Scarborough, London, and Calgary named for Sir Wilfrid, and primary schools in Vancouver, Brampton, Ont., and Ste. Rose du Lac, Man., among others named for him (although several spell his first name Wilfred).

One of Ottawa's most distinctive landmarks, the Château Laurier, was opened in June 1912 by Sir Wilfrid, then leader of the opposition. Across the Ottawa River in Hull, the Canadian Museum of Civilization is located in Laurier Park.

Remembering John F. Kennedy

Where were you on that black day in 1963 when the shots rang out in Dallas? For millions around the world, the answer is seared in our memories. It was as if a family member had been cut down at the height of his promise and, with him, the potential for a more caring society, a greater sense of dignity for the individual.

I heard the heart-rending news on that fateful 22 November in my office on Booth Street in Ottawa, where I was working as a toponymic researcher. I still recall joining with my colleagues to share our anguish and, later, tearfully comforting my wife, who, like myself, was deeply distressed by the assassination of President John F. Kennedy.

Almost immediately, President Lyndon B. Johnson proclaimed the renaming of Idlewild Airport at New York and Cape Canaveral Space Center and Cape Canaveral in Florida for the slain president. In 1973, the name Cape Canaveral was restored for the prominent point extending into the Atlantic Ocean, east of Orlando. The John F. Kennedy Space Center remains the name of the site for spacecraft launchings.

The earliest honour in Canada was Avenue du Président-Kennedy, given on 9 January 1964 by the city of Montréal for a new street south of McGill University, between and parallel to Sherbrooke and Sainte-Catherine. The province of Québec gave the name Route du Président-Kennedy to Highway 173 leading south from Lévis, on the St. Lawrence River opposite Québec City, to the Maine boundary.

In early 1964, the residents of Sevogle, N.B., 21 kilometres northwest of Newcastle, asked the provincial government to change the name of their small community to Mount Kennedy. A year later, the Northumberland County Council and the Miramichi Historical Society expressed their opposition. Nothing more was heard about Sevogle's efforts to sanctify – as I characterize commemorative community names like Mount Kennedy – the late president in the Miramichi.

For several months in 1963 I had been working on names proposed for mountain divisions for a new physiographic map of British Columbia. Specifically, I had been investigating the hierarchy of 'mountains,' 'ranges' within mountains, and each named 'range' within ranges.

In the northern Rockies, 160 kilometres west of Fort Nelson, Mount Roosevelt and Churchill Peak were found to be located in an unnamed range of the Muskwa Ranges. I suggested Statesmen Range for it, with the further suggestion that the unnamed peaks within it be reserved for prominent world statesmen. A few days after Kennedy's assassination I proposed commemorating him on one of the many unnamed peaks in the 'Statesmen Range.'

In early December 1963, the names of Statesmen Range and Mount Kennedy were submitted to the British Columbia representative on the Canadian Permanent Committee on Geographical Names, with the added suggestion that the selected peak might be visible from the Alaska Highway, 40 kilometres north of Mount Roosevelt.

The British Columbia government was not enthusiastic about either proposal, especially the idea of naming a group of peaks Statesmen Range. It did not like the idea of reserving peaks for noted international figures, preferring to have names evolve as the features were climbed by mountaineers or explored by geologists. With some reluctance, it selected a peak for Kennedy midway between Mount Roosevelt and Churchill Peak.

Later, another peak 10 kilometres northwest of Mount Roosevelt, and closer to the Alaska Highway, was selected. Prime Minister Lester B. Pearson wanted to make the announcement on the first anniversary of the assassination.

However, at its annual meeting on 5 October 1964, the Canadian Permanent Committee on Geographical Names, with convincing arguments put forward by its Alberta member (his province had never forgiven Mackenzie King for changing Castle Mountain to Mount Eisenhower in 1948), rejected the naming of a peak for Kennedy. It noted his position in world history could not yet be properly assessed, and pointed out that since few unnamed major peaks remained in Canada, they should be reserved for worthy Canadians.

But Pearson, a great admirer of the late charismatic president, was not to be dissuaded. Since the authority for naming in the northern territories was then vested in the federal government, the prime minister suggested a peak in the Yukon in the St. Elias Mountains, near the

Alaska boundary. An unnamed peak near Mount Logan was chosen, even though its elevation was found to be only 3,776 metres, making it more than 2,175 metres lower than Mount Logan, the highest point in Canada. On 20 November Pearson made the official announcement in the House of Commons.

Two weeks later, Walter Wood, then president of the American Geographical Society and a veteran mountain climber in Alaska and Yukon, informed the Canadian government that the peak named for the president was really only a part of the Mount Logan uplift. He suggested selecting one of the highest unnamed peaks in Canada, a distinct mountain 85 kilometres to the southeast, but within 6 kilometres of Mount Alverstone and Mount Hubbard on the United States–Canada boundary.

Bradford Washburn, then director of the Museum of Science in Boston, who had crossed the St. Elias Mountains in 1935 and photographed several lofty mountains *en route*, also urged Canadian authorities to adopt Wood's proposed peak. On 22 December, the new, higher Mount Kennedy was approved.

In March 1965, Sen. Robert F. Kennedy ascended the peak with seven experienced mountain climbers, and went on record as the first person to reach the 4,238-metre summit. Writing in *National Geographic* (July 1965), Senator Kennedy observed that his brother 'would be pleased that this lonely, beautiful mountain in the Yukon bears his name, and that in this way, at least, he has joined the fraternity of those who live outdoors, battle the elements, and climb mountains.' Five years later, Robert Kennedy also was gunned down, another victim of a violent decade for American public figures.

Half the presidents – twenty-one out of forty-two – have been honoured in one or more of Canada's geographical names. John F. Kennedy was the last to receive such an honour. With a policy now in place to limit commemorating of non-Canadians, it is unlikely other presidents will have their names assigned to geographical features in Canada.

Smith: We Are Blessed with 439 of Them!

No other family name is more common in the English-speaking world than Smith. In *The Book of Smiths* (1978), Elsdon Smith estimated that there were 2.8 million people with this surname in the world, with nearly 75,000 of them in Canada. It is the number-one surname in the United States, England, and Scotland, and is probably the leading surname in Canada as well.

Smith and its cognates Smyth and Smythe represent the most prominent family name among Canada's place names. The length and breadth of the country is blessed with 439 mountains, creeks, points, capes, rivers, and populated places with this surname. Not counted are such names as Arrowsmith, Goldsmith, or Sexsmith, but ones like Allansmith River and Ladysmith are included because Smith is a distinct component in the original family name.

Ontario has the largest number of Smith names, with ninety, followed by Newfoundland, with seventy-one; British Columbia, with fifty-eight; and Saskatchewan, with fifty-two. Prince Edward Island has no places or features called Smith; in 1765, Samuel Holland named Smith Point in Kings County for one of three officers he had served with, but today the name is unknown locally.

Few of the geographical names with Smith in Canada are particularly prominent, with perhaps only Smiths Falls in Ontario and Fort Smith in the Northwest Territories known fairly widely.

Smiths Falls commemorates Maj. Thomas Smyth, who

was granted 400 acres (162 hectares) at the Rideau River millsite in 1804, but subsequently lost his holdings in 1826 to Charles Jones, who sold out to Abel Ward. Ward, who built the first mills, called the new settlement Wardsville, but the settlers persisted in using Smyth's Falls. Smith's Falls was incorporated as a village in 1854, and as a town in 1882. Although it had long been spelled locally without the apostrophe, a special act of the Ontario legislature had to be passed in 1968 to allow the town to issue debentures in the name of Smiths Falls. A move in the 1930s to call it Rideau City – as a name more suitable for a rising metropolis – failed.

Fort Smith, on the N.W.T.–Alberta border, was established as a Hudson's Bay Company trading centre in 1874, and named for Sir Donald A. Smith, Lord Strathcona (1820–1914), a member of the first council of the Northwest Territories and president of the Canadian Pacific Railway. In 1910, Medicine Hat was almost renamed Smithville in his honour. Strathcona in Alberta, Mount Sir Donald in British Columbia, and Strathcona Sound in Nunavut were also named for him.

Smithville in the Niagara Peninsula was first known as Griffintown, but was renamed by Smith Griffin in the early 1800s for his mother's family. Foxboro near Belleville was called Smithville from about 1800 to about 1860, when the present name was adopted. A few miles to the southwest, near Trenton, is the community of Smithfield, where John Smith was a pioneer settler in 1793, and where Abijah Smith was the postmaster from 1860 to 1870.

On the west side of Peterborough is Smith Township, named in 1821 likely for Samuel Smith, then a member of the executive council of Upper Canada, who had been administrator of the province from 1817 to 1820, when a lieutenant-governor was not present. The township was probably not named after Sir David Smith (Smyth), surveyor general of Upper Canada, 1792–1804, who had returned to England in 1805.

The site of Port Hope, Ont., was settled in 1760 by Peter Smith, and became known as Smith's Creek. Subsequently, the place was called Toronto, and then Port Hope in 1817, and the stream was called Ganaraska River, a name of Aboriginal origin.

The name Smith Sound, suggesting an arm of the sea, occurs no fewer than four times in Canada. In 1616, William Baffin named Smith Sound, one of the narrow channels between Ellesmere Island and Greenland, for Sir Thomas Smith, an English merchant and promoter of exploration into the Northwest Passage. The origins of the same name in British Columbia, Nova Scotia, and Newfoundland are not noted in the records of the Geographical Names Board of Canada.

The naming of geographical features in the 1950s and 1960s for soldiers killed during the Second World War was widely practised by some provinces. Thus were forty features in Canada named Smith or a variant of it. Among Saskatchewan's fifty-two names with Smith are no fewer than five called Smith Peninsula, and four of them honour servicemen who lost their lives. Four features called Smith Lake in Saskatchewan also commemorate Second World War casualties. Smith-Jones Lake in that province was named in 1955 for Flt. Sgt. Henry Smith-Jones, who was killed in Europe in 1943; and Smith-Windsor Islands in Lake Athabasca was given in 1953 for Grenville Smith-Windsor, who was killed in North Africa in 1942.

Among British Columbia's fifty-eight names with Smith is Mount Robert Smith in the Selkirk Mountains, named for a land surveyor killed during the First World War. Mount DeCosmos, west of Nanaimo on Vancouver

Island, commemorates William Alexander Smith, but he was better known as Amor DeCosmos when he was premier of British Columbia and a federal MP in the 1870s.

Ladysmith occurs as a place name in Québec, Ontario, and British Columbia. They all commemorate the British victory at Ladysmith, Natal, in 1900. The latter was named in 1851 for the Spanish-born wife of the governor of Natal, Sir Harry Smith, who had rescued her during the sacking of Badajos in Spain in 1812, and later married her (see also pp. 123–8).

Unusual names with Smith across the country include Cougar Smith Creek in British Columbia, Canon Smith Lake in Alberta, Smith Coulee in Saskatchewan, Smiths Slough in Manitoba, Smiths Shanty Hill in Ontario, Lac Mackey-Smith in Québec, Billy Smith Rips in New Brunswick, Neddie Smiths Pond in Nova Scotia, and Hick Smith Stillwater in Newfoundland.

From Smiths Falls in Ontario to Fort Smith in the Northwest Territories, it is quite evident that the Smiths have had a lasting and distinctive impact on Canada's history.

Index